DESIGN OF EXPERIMENTS FOR GENERALIZED LINEAR MODELS

CHAPMAN & HALL/CRC
Interdisciplinary Statistics Series

Series editors: N. Keiding, B.J.T. Morgan, C.K. Wikle, P. van der Heijden

Recently Published Titles

CORRESPONDENCE ANALYSIS IN PRACTICE, THIRD EDITION
M. Greenacre

STATISTICS OF MEDICAL IMAGING
T. Lei

CAPTURE-RECAPTURE METHODS FORTHE SOCIAL AND MEDICAL SCIENCES
D. Böhning, P. G. M. van der Heijden, and J. Bunge

THE DATA BOOK
COLLECTION AND MANAGEMENT OF RESEARCH DATA
Meredith Zozus

MODERN DIRECTIONAL STATISTICS
C. Ley and T. Verdebout

SURVIVAL ANALYSIS WITH INTERVAL-CENSORED DATA
A PRACTICAL APPROACH WITH EXAMPLES IN R, SAS, AND BUGS
K. Bogaerts, A. Komarek, E. Lesaffre

STATISTICAL METHODS IN PSYCHIATRY AND RELATED FIELD
LONGITUDINAL, CLUSTERED AND OTHER REPEAT MEASURES DATA
Ralitza Gueorguieva

FLEXBILE IMPUTATION OF MISSING DATA, SECOND EDITION
Stef van Buuren

COMPOSITIONAL DATA ANALYSIS IN PRACTICE
Michael Greenacre

MODEL-BASED GEOSTATISTICS FOR GLOBAL PUBLIC HEALTH
METHODS and APPLICATIONS
Peter J. Diggle and Emanuele Giorgi

DESIGN OF EXPERIMENTS FOR GENERALIZED LINEAR MODELS
Kenneth G. Russell

For more information about this series, please visit: https://www.crcpress.com/go/ids

DESIGN OF EXPERIMENTS FOR GENERALIZED LINEAR MODELS

Kenneth G. Russell
National Institute for Applied Statistical Research Australia
University of Wollongong

CRC Press
Taylor & Francis Group
Boca Raton London New York

CRC Press is an imprint of the
Taylor & Francis Group, an **informa** business
A CHAPMAN & HALL BOOK

CRC Press
Taylor & Francis Group
6000 Broken Sound Parkway NW, Suite 300
Boca Raton, FL 33487-2742

International Standard Book Number-13: 978-1-4987-7313-3 (Hardback)

Library of Congress Cataloging-in-Publication Data

Names: Russell, K. G. (Kenneth Graham), author.
Title: Design of experiments for generalized linear models / K.G. Russell.
Description: Boca Raton, Florida : CRC Press, [2019] | Series: Chapman & Hall/CRC interdisciplinary statistics | Includes bibliographical references and index.
Identifiers: LCCN 2018041298| ISBN 9781498773133 (hardback : alk. paper) | ISBN 9781498773164 (e-book).
Subjects: LCSH: Linear models (Statistics) | Experimental design.
Classification: LCC QA276 .R875 2019 | DDC 519.5/7--dc23
LC record available at https://lccn.loc.gov/2018041298

Visit the Taylor & Francis Web site at
http://www.taylorandfrancis.com

and the CRC Press Web site at
http://www.crcpress.com

To Janet

Contents

Preface

Generalized linear models (GLMs) were introduced over 45 years ago. They combine the features of regression, analysis of variance (ANOVA) and Analysis of Covariance (ANCOVA), but allow these methods of analysis to be extended to a large number of distributions other than the normal distribution. Many books have been published that either are specifically devoted to GLMs or include a consideration of GLMs. If data have been collected that suit the application of a GLM, then you can probably find information on the features of a statistical analysis in a widely available book. However, there is very little information to tell you on how you should collect the data that are to be analysed in this way.

To the best of my knowledge, this is the first book to be written specifically on the design of experiments for GLMs. There are several handbooks or monographs that include a chapter or section on the topic. However, this material is usually brief, requires an advanced knowledge of mathematics, and provides little help with how one might actually write a computer program that will find an optimal design. My aim is to fill the gap by providing explanations of the motivation behind various techniques, reduce the difficulty of the mathematics or move it to one side if it cannot be avoided, and give lots of details on how to write and run computer programs using the popular statistical software R (R Core Team, 2018). I have also omitted some topics that regularly appear in other references. If I could not answer "yes" to the question, "Well, that's nice — but is it useful?", the topic was omitted.

This book provides an introduction to the theory of designing experiments for GLMs. It then looks in depth at applications for the binomial and Poisson distributions. Smaller segments consider the multinomial and gamma distributions and situations where a specific distribution is not assumed (quasi-likelihood methods). The final chapter considers Bayesian experimental designs.

While mathematics provides the fundamental underpinning of the design of experiments for GLMs, the mathematics required to follow this book has deliberately been kept light. I am very aware that many researchers in the sciences and social sciences do not have an extensive mathematics training. I assume that the reader has done elementary differentiation and understands why we are interested in finding a gradient when seeking to maximise or minimise a function. Some minimal matrix algebra is also required: knowing what a matrix is, understanding the concept of a matrix inverse, and being able to multiply together two compatible matrices and to find the transpose of a matrix. Of course it will be helpful if your mathematical skills are deeper than this (there is less that you will have to take for granted), but you should find that anything more advanced than this is explained or illustrated.

The R software package is very widely used for statistical work. Probably its most attractive feature for many users is that it is free, but it would be a serious mistake to assume that this is its only attraction. New functions are being added to R all the time. There is a comprehensive literature available to help use it, and it is easy to obtain help through chat sites and other online facilities. I have chosen to use this software because of its accessibility by all potential designers of experiments. I have incorporated R programs in numerous examples in the text, and have included a set of R programs in a Web site (doeforglm.com) available to all readers. This is *not* an R package. I make no claim for elegance in my usage of R. For example, the use of loops is frowned upon by R specialists, but I find it much easier to understand what a program is doing when I see a loop than when I see a shortcut. I have aimed for good exposition rather than optimised computing. However, loops have been replaced in a number of places after the purpose of the loop has been demonstrated.

I have not attempted to give a detailed Bibliography that covers all research on the design of experiments for GLMs. I have restricted references to those that are essential and (ideally) easily available. Going through the reference lists provided in these references should provide the interested reader with a near-complete list of relevant material. I apologise to any authors who feel slighted by the omission of their work from this abbreviated Bibliography.

I hope that this book will be of use to researchers who want to run efficient designs that will collect data to be analysed by GLMs. Professional statisticians should find that there is enough appropriate material here to take them to the borders of new statistical research. I trust that nonstatisticians will find enough here for them to be able to design their own experiments, by following the examples and using the programs provided.

I would like to thank the many researchers who have consulted me over the years. They made it very obvious that a lack of deep mathematical understanding does not coincide with a lack of research ability. If I have got the mathematics, explanations and computer programs at the right level for you, I will consider myself well satisfied. I would also like to thank my experimental design colleagues for their assistance in giving me an understanding of the design of experiments for GLMs. I am indebted to the reviewers of the drafts of this manuscript for their helpful comments and constructive criticisms. Any faults that remain are my responsibility alone. I extend my sincere thanks to the editorial and production staff at CRC Press for their superb support during the writing and production of this book. Finally, I thank the many people who have posted answers to questions about LaTeX or R on websites. Many was the time that I wondered how to do something in LaTeX or R and found the answer already on the Web.

Kenneth G. Russell

Chapter 1

Generalized Linear Models

1.1 Introduction

Numerous books exist on the subject of generalized linear models (GLMs). They tell you how to analyse data that have already been collected. Unlike those books, this one does not consider data analysis, but focusses on how you should design the experiment that will collect the data. Designing an experiment is beneficial, as it can increase the information that you obtain from your experiment without increasing the resources that are required.

Before discussing GLMs, I briefly review the linear models that are being generalised. The emphasis is on features of linear models that are of importance for GLMs.

Regression analysis and the analysis of variance (ANOVA) are important techniques in the statistical toolkits of most users of statistics. Although often taught as separate procedures, they both belong within the topic of linear models. These methods of analysis are often used in experimental situations where a researcher takes several independent observations on a variable of interest, the *response variable* Y. This quantity is considered to come from a normal distribution with an unknown mean, μ, and an unknown standard deviation, σ. Then one or more *explanatory* or *predictor variables* thought to affect Y are changed in value, and several independent observations are made on Y again. This time the observations are considered to come from a normal distribution with a different (unknown) mean but the same unknown standard deviation.

Regression analysis and ANOVA are used to assess a mathematical model that purports to explain the variation observed in Y. The mathematical model consists of two parts:

- the first seeks to explain the changes in the population mean, μ, as the values of the explanatory variables are altered,
- the second aims to represent the variability of individual observations around their means.

The part of the model that explains changes in μ involves a function of the explanatory variables, and generally contains several parameters.

For example, if it is thought that the relationship between μ and an explanatory variable, x, is a straight line, this might be written as $\mu = \beta_0 + \beta_1 x$, where β_0 (the intercept of the line) and β_1 (the line's slope) are the two parameters of the model. They are unknown values that we attempt to estimate from the collected data.

Note that there are two types of parameters being considered here. The first type is the population parameter whose behaviour we wish to model. The other type is the collection of those parameters that are in the function used to explain the population parameter. In this book, the first type will be described as *population parameters*, or distribution parameters, while the second type will be known as *model parameters.* In the example above, μ is a population parameter, and β_0 and β_1 are model parameters.

Linear regression and ANOVA have the special requirement that the function of the explanatory variables that is used to model μ must be *a linear combination of the parameters of the model.* This is what makes them *linear* models. A "linear combination of the parameters of the model" will now be defined.

Assume that there are p model parameters, denoted here by $\theta_1, \ldots, \theta_p$. (When appropriate, other symbols will subsequently be used for the parameters.) Then

$$\eta = a_1\theta_1 + a_2\theta_2 + \cdots + a_p\theta_p \tag{1.1}$$

is a linear combination of the parameters if the multipliers (or "coefficients") $a_1, \ldots, a_p$ have known numerical values. Thus $\theta_1 = 1 \times \theta_1$, $\theta_1 + \theta_2 = 1 \times \theta_1 + 1 \times \theta_2$, and $2\theta_3 - \theta_4 - \theta_5 = 0 \times \theta_1 + 0 \times \theta_2 + 2 \times \theta_3 + (-1) \times \theta_4 + (-1) \times \theta_5$ are linear combinations of the parameters. However, $\theta_1 \times \theta_2$ is *not* a linear combination of the parameters, because the multiplier of θ_2, namely θ_1, is a parameter, not a known numerical value. The known numerical values $a_1, \ldots, a_p$ will come from values of the explanatory variables.

In linear regression and ANOVA, it is customary to include expressions for both the mean and the variability around the mean in a model for the response variable. The mathematical model may be written as

$$\begin{aligned} \text{response} &= \text{(linear combination of parameters)} + \text{random error} \\ &= \eta + \text{random error}. \end{aligned} \tag{1.2}$$

The *random error* is intended to explain why two observations on individuals with the same values of the explanatory variables do not equal one another. It collects together variation from natural causes, recording errors, a failure to include all relevant variables in the set of explanatory variables, etc.

In early applications of regression analysis and ANOVA, the main distinction between the two methods was the nature of the explanatory variables in the model:

- in regression analysis, their values were real numbers;
- in ANOVA, they were "indicator" variables (containing 0s and 1s to link the appropriate model parameters to the relevant observations).

Example 1.1.1. *Box, Hunter, & Hunter (2005, pp. 381–382) considered data from an investigation on the growth rates of rats that were fed various doses of a dietary supplement. It had been expected that, over the range of the supplement that had been used in the experiment, there would be a straight line relationship between the growth rate and the dose. However, a plot of the data suggested a curve, and so a quadratic model was fitted instead. The model was*

$$Y_i = \beta_0 + \beta_1 x_i + \beta_2 x_i^2 + E_i \qquad (i = 1, \ldots, 10), \tag{1.3}$$

where Y_i represents the growth rate observed on the ith rat, which had received an amount x_i of the supplement, and E_i represents the random error associated with the ith observation. By convention, the parameters in this model are denoted by β_0, β_1 and β_2 rather than the θ_1, θ_2 and θ_3 of (1.1).

The expression

$$\eta = \beta_0 + \beta_1 x_i + \beta_2 x_i^2 = 1 \times \beta_0 + x_i \times \beta_1 + x_i^2 \times \beta_2$$

is clearly a linear combination of the parameters β_0, β_1 and β_2. For $\beta_2 > 0$, it takes the shape (concave upwards) illustrated in Figure 1.1; if $\beta_2 < 0$, we would obtain a similar curve, but concave downwards. The actual shape that is seen depends on what domain of x is considered. If that domain lay entirely to the right of the minimum of the curve in Figure 1.1, then one would see only an upward-tending curve.

Important note: It is essential to remember that the word "linear" refers to a linear combination of the parameters, and does not mean that the model predicts a straight line relationship between Y and x. The model in Equation (1.3) is a linear model, even though the relationship between Y and x is a curve, not a straight line.

Example 1.1.2. *Kuehl (2000, pp. 264–265) described a randomised complete block design in which several different timing schedules of applications of nitrogen fertilizer were to be compared for their effect on the observed nitrate content from a sample of wheat stems. These timing schedules represented the "treatments" (six in all). As the experiment*

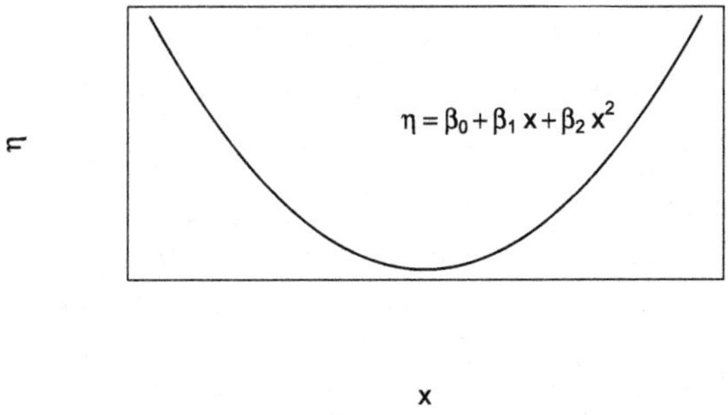

Figure 1.1 *The quadratic curve* $\eta = \beta_0 + \beta_1 x + \beta_2 x^2$ *for* $\beta_2 > 0$.

was conducted in a field with a water gradient arising from irrigation, the field was partitioned into sections that were relatively homogeneous with regard to moisture. These sections formed the "blocks" (four in all). Each treatment was randomly allocated to one of six plots within each block. This ensured that each treatment appeared in both the "wetter" and "dryer" parts of the field, and was not advantaged or disadvantaged by being randomly allocated to plots with just one or two water levels.

The statistical model for the response variable (observed nitrate content from a sample of wheat stems), Y_{ij}*, from the plot in block* j *that received treatment* i *was*

$$Y_{ij} = \mu + \tau_i + \rho_j + E_{ij} = \eta_{ij} + E_{ij}, \qquad i = 1, \ldots, 6;\ j = 1, \ldots, 4, \quad (1.4)$$

where μ *is the overall mean yield for all possible timings in all possible blocks,* τ_i *is the effect of the* i*th timing,* ρ_j *is the effect of the* j*th block, and* E_{ij} *represents the random error term. It is convenient to write* η_{ij} *for the linear combination of parameters that is used to model* μ_{ij}*, the mean of* Y_{ij}*.*

There are 11 model parameters in (1.4)*:* μ*,* $\tau_1, \ldots, \tau_6$ *and* $\rho_1, \ldots, \rho_4$*. These take the place of* $\theta_1, \ldots, \theta_{11}$ *in* (1.1)*. When there are different sources of variation, it is customary to use different symbols (e.g.,* τ *and* β*) for those different sources, rather than just the* $\theta_1, \theta_2, \ldots$ *of* (1.1)*.*

Although it is not immediately obvious, each parameter in (1.4) *is multiplied by the value of an indicator variable, and the parameters in the right-hand side of the model are a linear combination of all 11 parameters. For example, the observation on the plot in the first block that received treatment 1 may be written as*

$$Y_{11} = \mu + \tau_1 + \rho_1 + E_{11} = \eta_{11} + E_{11},$$

where

$$\eta_{11} = 1 \times \mu + 1 \times \tau_1 + 0 \times \tau_2 + \cdots + 0 \times \tau_6 + 1 \times \rho_1 + 0 \times \rho_2 + \cdots + 0 \times \rho_4.$$

The 0s and 1s which are multiplied by the parameters are values of the indicator variables.

Example 1.1.3. *The Michaelis-Menten model is often used in biological situations to explain the behaviour of a response variable Y in terms of an explanatory variable x. It has two parameters, β_1 and β_2. The model says*

$$Y_i = \frac{\beta_1 x_i}{1 + \beta_2 x_i} + E_i \qquad (i = 1, \ldots, n).$$

The function that models the mean, μ, is $\beta_1 x/(1 + \beta_2 x)$, but this cannot be written in the form $a_1\beta_1 + a_2\beta_2$, so the Michaelis-Menten model is not a linear model.

Earlier examples described situations where the explanatory variables in a linear model were all real variables (Example 1.1.1) or were all indicator variables (Example 1.1.2). However, in some experimental situations, the explanatory variables in a linear model would include both categorical or ordinal variables *and* also "real" variables. The latter variables were known as covariates, and the analysis of such linear models was called analysis of covariance (ANCOVA).

Example 1.1.4. *Kuehl (2000, pp. 551–553) described an experiment which examined the effects of exercise on oxygen ventilation. Twelve healthy males were recruited, and six each were allocated randomly to two exercise regimes (the two "treatments"). The response variable was the change in maximal oxygen uptake from the beginning to the end of the training period. The aim of the experiment was to compare the effects of the two treatments on this variable. As it was thought that the age of a subject might influence the response variable, age was also included in the model. Let Y_{ij} denote the change in maximal oxygen uptake of the jth subject in the ith group. Then the equation that was used to model the variation in the response variable was*

$$Y_{ij} = \eta_{ij} + E_{ij} = \mu_i + \beta_i(x_{ij} - \bar{x}_{..}) + E_{ij}, \quad i = 1, 2;\ j = 1, \ldots, 6, \quad (1.5)$$

where μ_i represents the mean effect of the ith treatment, x_{ij} is the age of the jth person in the ith treatment group, $\bar{x}_{..}$ is the average age of

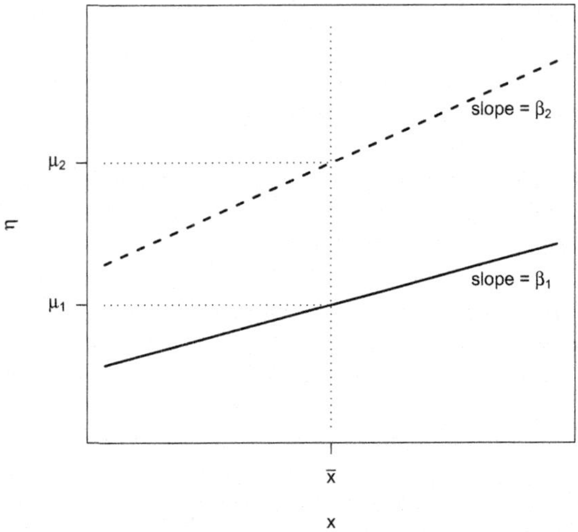

Figure 1.2 *One possible situation for the ANCOVA model in* (1.5).

all 12 subjects, and E_{ij} *represents the random error associated with the observation* Y_{ij}.

Equation (1.5) *suggests that the points* (x_{ij}, η_{ij}) *lie around one of two lines. The line for treatment* i $(i = 1, 2)$ *has an* η*-value of* μ_i *when* $x = \bar{x}_{..}$, *and has a slope of* β_i. *Figure 1.2 provides one possible example of this situation.*

The ANCOVA model in (1.5) *is a linear model involving four model parameters:* μ_1, μ_2, β_1 *and* β_2. *For the observation on (say) the third person in treatment group 2, the model says*

$$\begin{aligned} Y_{23} &= \eta_{23} + E_{23} \\ &= \mu_2 + \beta_2(x_{23} - \bar{x}_{..}) + E_{23} \\ &= 0 \times \mu_1 + 1 \times \mu_2 + 0 \times (x_{13} - \bar{x}_{..}) \times \beta_1 + 1 \times (x_{23} - \bar{x}_{..}) \times \beta_2 + E_{23}. \end{aligned}$$

In the linear combination of the four parameters, the coefficients are indicator variables or products of indicator variables with real variables.

In earlier years, when statistical analyses were done by hand, or with a calculator, ANOVA, ANCOVA and regression analysis were treated as separate types of analyses, to take advantage of simplifications in the calculations that arose as a result of this separation. However, with the advent of sophisticated statistical computing packages, it became possible to reduce the three methods of analysis to a single common method, which is often called the *general linear model.* Regrettably, this has sometimes been abbreviated as "GLM", which is also the acronym of the generalized linear models that are the subject of this book. In this book, the acronym "GLM" will be reserved for generalized linear models, and "general linear models" will always be written in full.

1.2 Mathematics of the general linear model

In commonly used notation for a general linear model, a regression model involving $(p-1)$ explanatory variables $x_1, \ldots, x_{p-1}$ is written as

$$Y_i = \beta_0 + \beta_1 x_{i1} + \beta_2 x_{i2} + \cdots + \beta_{p-1} x_{i,p-1} + E_i \quad (i = 1, \ldots, N),$$

where N is the total number of observations that are made on the response variable. There are p model parameters, $\beta_0, \ldots, \beta_{p-1}$. For notational convenience, the N model equations

$$\begin{aligned} Y_1 &= \beta_0 + \beta_1 x_{11} + \beta_2 x_{12} + \cdots + \beta_{p-1} x_{1,p-1} + E_1 \\ Y_2 &= \beta_0 + \beta_1 x_{21} + \beta_2 x_{22} + \cdots + \beta_{p-1} x_{2,p-1} + E_2 \\ &\vdots \\ Y_N &= \beta_0 + \beta_1 x_{N1} + \beta_2 x_{N2} + \cdots + \beta_{p-1} x_{N,p-1} + E_N \end{aligned}$$

are written in matrix and vector form as

$$\begin{bmatrix} Y_1 \\ Y_2 \\ \vdots \\ Y_N \end{bmatrix} = \begin{bmatrix} 1 & x_{11} & x_{12} & \cdots & x_{1,p-1} \\ 1 & x_{21} & x_{22} & \cdots & x_{2,p-1} \\ \vdots & \vdots & \vdots & \cdots & \vdots \\ 1 & x_{N1} & x_{N2} & \cdots & x_{N,p-1} \end{bmatrix} \begin{bmatrix} \beta_0 \\ \beta_1 \\ \vdots \\ \beta_{p-1} \end{bmatrix} + \begin{bmatrix} E_1 \\ E_2 \\ \vdots \\ E_N \end{bmatrix},$$

or as

$$\boldsymbol{Y} = \boldsymbol{X}\boldsymbol{\beta} + \boldsymbol{E}. \tag{1.6}$$

The $N \times p$ matrix $\boldsymbol{X}$ is frequently called the *design matrix.*

If $\boldsymbol{A}$ is an $m \times n$ matrix whose (i, j) element is a_{ij}, the *transpose* of $\boldsymbol{A}$ is the $n \times m$ matrix whose (i, j) element is a_{ji}. It is denoted by $\boldsymbol{A}^\top$ in this book.

Estimation of the value of $\boldsymbol{\beta}$ is a major aim of most experiments. Two common methods of estimating $\boldsymbol{\beta}$ in a general linear model are least squares (LS) estimation and maximum likelihood (ML) estimation. To prevent a break in continuity of exposition, details of these two methods will be deferred to Section 1.5.

It is usually assumed that the error terms are independent of one another and that each has a $N(0, \sigma^2)$ distribution. If these statistical assumptions are satisfied, then the LS and ML estimators of $\boldsymbol{\beta}$, $\hat{\boldsymbol{\beta}} = (\hat{\beta}_0, \hat{\beta}_1, \ldots, \hat{\beta}_{p-1})^\top$, are identical and can be shown (e.g., Guttman, 1982, Chapter 3 or Searle, 1971, Chapter 3) to satisfy the so-called *normal equations*

$$\boldsymbol{X}^\top \boldsymbol{X} \hat{\boldsymbol{\beta}} = \boldsymbol{X}^\top \boldsymbol{Y}. \tag{1.7}$$

In the regression context, the columns of $\boldsymbol{X}$ are usually of full column rank, which means that no column of $\boldsymbol{X}$ can be written as a linear combination of other columns. In this case, the $p \times p$ matrix $\boldsymbol{X}^\top \boldsymbol{X}$ is nonsingular and has a proper inverse, $(\boldsymbol{X}^\top \boldsymbol{X})^{-1}$. Then $\hat{\boldsymbol{\beta}}$ is given by

$$\hat{\boldsymbol{\beta}} = (\boldsymbol{X}^\top \boldsymbol{X})^{-1} \boldsymbol{X}^\top \boldsymbol{Y}.$$

It can be shown that

$$\mathrm{E}(\hat{\boldsymbol{\beta}}) = \boldsymbol{\beta}, \tag{1.8}$$

and that the covariance matrix of $\hat{\boldsymbol{\beta}}$ is equal to

$$\mathrm{cov}(\hat{\boldsymbol{\beta}}) = \sigma^2 (\boldsymbol{X}^\top \boldsymbol{X})^{-1}. \tag{1.9}$$

Example 1.2.1. *For the straight line regression model*

$$Y_i = \beta_0 + \beta_1 x_i + E_i \quad (i = 1, \ldots, N),$$

where the single explanatory variable is labelled x for simplicity, there are two model parameters, β_0 and β_1, and the $N \times 2$ matrix $\boldsymbol{X}$ satisfies

$$\boldsymbol{X} = \begin{bmatrix} 1 & x_1 \\ 1 & x_2 \\ \vdots & \vdots \\ 1 & x_N \end{bmatrix}.$$

It can be shown that

$$\boldsymbol{X}^\top \boldsymbol{X} = \begin{bmatrix} N & \sum_{i=1}^N x_i \\ \sum_{i=1}^N x_i & \sum_{i=1}^N x_i^2 \end{bmatrix}$$

and

$$\sigma^2 (\boldsymbol{X}^\top \boldsymbol{X})^{-1} = \frac{\sigma^2}{\Delta} \begin{bmatrix} \sum_{i=1}^N x_i^2 & -\sum_{i=1}^N x_i \\ -\sum_{i=1}^N x_i & N \end{bmatrix}, \tag{1.10}$$

where

$$\Delta = N\sum_{i=1}^{N} x_i^2 - (\sum_{i=1}^{N} x_i)^2.$$

The variance of $\hat{\beta}_1$ is given by the $(2,2)$ element of $\sigma^2(\boldsymbol{X}^\top\boldsymbol{X})^{-1}$. Using the standard result that

$$N\sum_{i=1}^{N} x_i^2 - (\sum_{i=1}^{N} x_i)^2 = N\sum_{i=1}^{N}(x_i - \bar{x})^2, \tag{1.11}$$

it follows that

$$var(\hat{\beta}_1) = \frac{\sigma^2}{N\sum_{i=1}^{N}(x_i - \bar{x})^2} \times N = \frac{\sigma^2}{\sum_{i=1}^{N}(x_i - \bar{x})^2}.$$

In an ANOVA context, the ML estimator of the vector of model parameters still satisfies (1.7), but the columns of $\boldsymbol{X}$ do not have full column rank under commonly used statistical models, and $\boldsymbol{X}^\top\boldsymbol{X}$ is singular. The matrix $(\boldsymbol{X}^\top\boldsymbol{X})^{-1}$ does not exist, and there is not a unique value of $\hat{\boldsymbol{\beta}}$. This is not the insurmountable problem that it might seem. Consider Example 1.1.2, in which several different timing schedules of applications of nitrogen fertilizer (the treatments) were to be compared. Although the estimates of the different treatment effects $\hat{\tau}_i$ are not unique, the estimates of the differences between *pairs* of treatment effects, $\hat{\tau}_i - \hat{\tau}_j$, *are* unique in this experiment. For example, see Searle (1971, Chapter 5). This is what is required for comparisons of treatment effects.

1.3 Towards the generalized linear model

The general linear model provides various sub-models that have been useful in many contexts. Under the assumptions that the response variable is normally distributed with constant variance, there is well-developed theory for the estimation of the model parameters and the testing of hypotheses about appropriate functions of these parameters. This theory is widely available in many books at various levels of sophistication. An additional advantage is that, when the parameters are regarded as fixed constants, the computation required to perform an analysis is exact and straightforward.

Probably for these reasons, the use of the general linear model in statistical analyses continues to be very popular. This is eminently appropriate when the assumptions underlying the general linear model have been met (at least to a good approximation). However, if the assumptions are violated, then the use of the general linear model becomes problematic. In earlier days, it was common to introduce a transformation of the response variable, and to analyse the transformed variable in place of the original variable (e.g., to use $\sqrt{Y}$ as the response variable instead of Y). If the transformation was applied to data that were not approximately normally distributed, and the new variable was much closer to being "normal", and/or if the transformation had the effect of making the data have essentially a common variance, then it would ease the researcher's task considerably if standard procedures for the general linear model could be applied to transformed data.

However, selected transformations do not always produce data that are both normally-distributed *and* have a constant variance. At other times, no sensible transformation could be found that met either of these aims. It was clear that alternative methods of analysis were needed.

One approach is not to transform the response variable, but instead to transform the population parameter that is of interest. In the general linear model, it is μ that we wish to model. So the function of the population parameter, $\mu = \mathrm{E}(Y)$, that we are seeking to estimate is just the mean itself. Mathematicians say that we are estimating the *identity function* of μ: $g(\mu) = \mu$ (the function that maps any entity onto itself).

If the general linear model is viewed as the first example of transforming the distribution parameter to something convenient, then the particular function of μ that is modelled by η is the identity function.

This approach should not cause concern. A linear combination of parameters,

$$\eta = a_1\theta_1 + a_2\theta_2 + \cdots + a_p\theta_p,$$

can potentially take any possible value (positive, negative or zero), as can the value, μ, of the mean. So the value of the linear combination is not required to be equal to something that is not possible for the value of μ. There is no mathematical difficulty here.

You may wish to object that the mean weight of (say) the wheat harvested from a field cannot be negative, and so the mean should not be modelled by a linear combination that can take negative values. You are correct. However, as stated by Box, Hunter, & Hunter (2005, p. 440), "*The most that can be expected from any model is that it can supply a useful approximation to reality: All models are wrong; some models are useful.*" The fact that a negative mean might be predicted for some values of the explanatory variable(s) indicates a deficiency in the model *for this particular example*, but it does not rule out the use of a linear combination of parameters to predict the mean.

A more important difficulty arises when one models a population parameter that *always* has strong restrictions on its possible values. Consider the probability, π, that a particular outcome occurs when a phenomenon is observed. This may be the probability that a rat is killed by the administration of a particular dose, x, of a poison. It is reasonable to expect the value of π to vary as x is altered, and consequently one might wish to model π by a linear model involving x. However, the value of π must lie between 0 and 1, not just for this problem but for all situations involving a probability. So π should not be modelled by a linear combination that takes values outside this interval. Rather than restrict the value of the linear combination, we look for some function of π, $g(\pi)$, that can take any value at all, and model that function by the linear combination. One such function is

$$g(\pi) = \ln\left(\frac{\pi}{1-\pi}\right) \qquad (0 < \pi < 1),$$

which is known as the *logit* function. As $0 < \pi < 1$, then $0 < 1 - \pi < 1$ also, and so $\pi/(1-\pi) > 0$. Then $\ln[\pi/(1-\pi)] < 0$ for $0 < \pi/(1-\pi) < 1$ and $\ln[\pi/(1-\pi)] \geq 0$ for $\pi/(1-\pi) \geq 1$. So $\ln[\pi/(1-\pi)]$ can be modelled by a linear combination that may take any positive or negative value.

Another population parameter with a restricted domain is the rate of occurrence (often denoted by λ) in a Poisson distribution. The rate cannot be negative, so $\lambda \geq 0$. This implies that $\ln(\lambda) < 0$ for $0 \leq \lambda < 1$, and $\ln(\lambda) \geq 0$ for $\lambda \geq 1$. Thus the natural logarithm function is a possible function of λ that might be modelled by a linear combination of parameters that takes positive or negative values.

Irrespective of the distribution of interest, the function of the population parameter θ that is modelled by a linear combination of model parameters is typically denoted by $g(\theta)$, and is called the *link function.*

In a normal distribution, the variance, σ^2, is not functionally related to the mean, μ. In some special circumstance, one might define the variance to be (say) μ^2, but this would be for a reason related to that situation. There is no general link between μ and σ^2. However, this lack of a relationship between the mean and variance is peculiar to the normal distribution. For other distributions, there is a formal relationship between the two. It is common to write

$$\text{var}(Y) = \phi V(\mu), \tag{1.12}$$

where ϕ is a constant and $V(\mu)$ is a function of μ. For the normal distribution, $\phi = \sigma^2$ and $V(\mu) = 1$.

Consider the *Poisson* distribution, where the population variance is equal to the population mean. So $\text{var}(Y) = \phi V(\mu)$, where $\phi = 1$ and

$V(\mu) = \mu$. Alternatively, consider the *binomial* distribution, whose distribution parameters are n (a known positive integer) and π (the probability of "success"). The mean of the distribution is $\mu = n\pi$, and the variance is $n\pi(1-\pi)$, so there is clearly a functional relationship between them: $\text{var}(Y) = \phi V(\mu)$ where $\phi = 1$ and $V(\mu) = \mu(1-\mu/n)$. The *Bernoulli* distribution is a binomial distribution with $n = 1$; for this distribution, $\text{var}(Y) = \phi V(\mu)$ where $\phi = 1$ and $V(\mu) = \mu(1-\mu)$.

If one seeks to extend the general linear model to situations beyond the standard normal distribution, one needs to know what the functional relationship is between the variance and the mean of the distribution.

An additional requirement is to know the particular distribution from which data are being taken. Knowledge of this gives a greater mathematical structure on which one can base the estimation of the model parameters, or the testing of hypotheses about their values.

As outlined above, there are three requirements of a model:

1. An appropriate function (the 'link' function) of a population parameter can be modelled by a linear combination of model parameters.
2. The response variable has a known distribution.
3. The functional relationship between the variance and the mean of this distribution is known.

When these three requirements are met, a GLM can be formed. It is possible to relax requirement 2; see Section 6.4.

1.4 Generalized linear models

Generalized linear models were introduced in Nelder & Wedderburn (1972). The definitive reference is McCullagh & Nelder (1989), but there are now numerous books that deal, at least in part, with GLMs. Some of these books (e.g., Faraway, 2006) address the analysis of GLMs with the statistical package R (R Core Team, 2018), which will be used in this book to seek optimal designs. These models have considerably increased the number of distributions from which data can be analyzed without violating underlying assumptions. Additionally, they have provided an umbrella for various methods of analysis that were previously considered to be unrelated, such as logistic regression, probit regression and loglinear regression.

A readable introduction to the theory of GLMs is given in Dobson & Barnett (2008) where, on page 46, the authors consider a single random variable Y whose probability (density) function depends on a single population parameter, θ. If the function depends on more than one population parameter (e.g., the normal distribution's density function, which depends on both μ and σ), the second parameter is treated as a constant. The distribution is said to belong to the *exponential family of distributions* if the probability (density) function of Y can be written in the form

$$f_Y(y;\theta) = \exp[a(y)b(\theta) + c(\theta) + d(y)]. \tag{1.13}$$

Other expressions exist for the form of probability functions of distributions belonging to the exponential family, but they are all equivalent.

Example 1.4.1. *The binomial probability function is commonly written as*

$$f_Y(y;\pi) = \binom{n}{y} \pi^y (1-\pi)^{n-y}, \quad y = 0, 1, \ldots, n;\ 0 < \pi < 1.$$

While it has two population parameters, n (the number of trials) and π (the probability of success on an individual trial), the value of n is generally known, and π is regarded as the population parameter of interest. We may write

$$\begin{aligned} f_Y(y;\pi) &= \binom{n}{y} \pi^y (1-\pi)^{n-y} \\ &= \exp[\ln \binom{n}{y} + y \ln \pi + (n-y)\ln(1-\pi)] \\ &= \exp\{y[\ln \pi - \ln(1-\pi)] + n\ln(1-\pi) + \ln \binom{n}{y}\} \\ &= \exp[a(y)b(\pi) + c(\pi) + d(y)], \end{aligned}$$

where $a(y) = y$, $b(\pi) = \ln \pi - \ln(1-\pi) = \ln[\pi/(1-\pi)]$, $c(\pi) = n\ln(1-\pi)$ and $d(y) = \ln \binom{n}{y}$. So the binomial probability function takes the form given in (1.13), and therefore the binomial distribution is a member of the exponential family of distributions.

Example 1.4.2. *The normal distribution with mean μ and variance σ^2, commonly denoted by $N(\mu, \sigma^2)$, also depends on two population parameters, μ and σ, but interest is generally in estimating μ. The quantity σ is regarded as a "nuisance parameter" and treated as a constant when examining whether the normal distribution belongs to the exponential family. Given $Y \sim N(\mu, \sigma^2)$, the probability density function of Y may be written as*

$$\begin{aligned} f_Y(y;\mu) &= 1/\sqrt{(2\pi\sigma^2)} \exp[-(y-\mu)^2/(2\sigma^2)] \\ &= \exp[-0.5\ln(2\pi\sigma^2) - (y^2 - 2y\mu + \mu^2)/(2\sigma^2)] \\ &= \exp\{y(\mu/\sigma^2) - \mu^2/(2\sigma^2) - [0.5\ln(2\pi\sigma^2) + y^2/(2\sigma^2)]\} \\ &= \exp[a(y)b(\mu) + c(\mu) + d(y)], \end{aligned}$$

Distribution	θ	$a(y)$	$b(\theta)$	$c(\theta)$	$d(y)$
Binomial (n, π) n known	π	y	$\ln[\pi/(1-\pi)]$	$n \ln(1-\pi)$	$\ln \binom{n}{y}$
Poisson (λ)	λ	y	$\ln \lambda$	$-\lambda$	$-\ln(y!)$
Normal (μ, σ^2) σ known	μ	y	μ/σ^2	$\mu^2/(2\sigma^2)$	$-[\ln(2\pi\sigma^2) + y^2/\sigma^2]/2$
Gamma (α, β) α known	β	y	$-\beta$	$\alpha \ln \beta$	$(\alpha - 1)\ln y - \ln \Gamma(\alpha)$

Table 1.1 *Some distributions that are members of the exponential family of distributions.*

where $a(y) = y$, $b(\mu) = \mu/\sigma^2$, $c(\mu) = -\mu^2/(2\sigma^2)$ *and* $d(y) = -[\ln(2\pi\sigma^2) + y^2/\sigma^2]/2$. *As* $f_Y(y;\mu)$ *has the form given in* (1.13), *the normal distribution belongs to the exponential family of distributions.*

Example 1.4.3. *The gamma distribution with a scale parameter,* β, *and a shape parameter,* α, *has probability distribution*

$$f_Y(y) = \frac{\beta^\alpha}{\Gamma(\alpha)}\, y^{\alpha-1} \exp(-\beta y), \quad y > 0 \quad (\alpha > 0,\, \beta > 0), \tag{1.14}$$

where $\Gamma(\cdot)$ *is the gamma function. The shape parameter of the gamma distribution is considered known, and the scale parameter* β *is the parameter in which we are interested. It can easily be shown that the gamma distribution belongs to the exponential family, and that* $E(Y) = \mu = \alpha/\beta$ *and* $var(Y) = \alpha/\beta^2 = (1/\alpha)\mu^2$.

Table 1.1 displays some commonly used distributions belonging to the exponential family.

If Y is an observation from a distribution belonging to the exponential family then, from Dobson & Barnett (2008, p. 49),

$$E[a(Y)] = -\frac{c'(\theta)}{b'(\theta)} \quad \text{and} \quad \text{var}[a(Y)] = \frac{b''(\theta)c'(\theta) - c''(\theta)b'(\theta)}{[b'(\theta)]^3}, \tag{1.15}$$

where $b'(\theta)$ and $b''(\theta)$ represent, respectively, the first and second derivatives of $b(\theta)$ with respect to θ.

When $a(y) = y$, the distribution is said to have the *canonical* ("standard") form. For distributions of this form, (1.15) provides expressions for E(Y) and var(Y). Of more importance to this book, for distributions having the canonical form, we can use a GLM to perform statistical inference on the model parameters in the linear combination used to model the link function, $g(\theta)$, of the population parameter θ. When the distribution has the canonical form, the function $b(\theta)$ is said to be the *canonical link function.*

It is usual to denote the p model parameters in the linear combination by $\beta_0, \ldots, \beta_{p-1}$. That is, we write

$$g(\theta) = \eta = \beta_0 + \beta_1 x_1 + \cdots + \beta_{p-1} x_{p-1}.$$

Recall from Section 1.2 that the coefficients of the model parameters in η for the ith observation form the ith row of the design matrix $\mathbf{X}$.

Although the notation $\boldsymbol{X}$ for the design matrix is fairly standard in regression analysis, it is less often used when it comes to designing experiments, especially for GLMs. We shall frequently be interested in maximising or minimising some function (e.g., the determinant – see page 26) of a matrix of the form $\boldsymbol{X}^\top \boldsymbol{W} \boldsymbol{X}$, where $\boldsymbol{W}$ is (usually) a diagonal matrix. This will require choosing the rows of $\boldsymbol{X}$. The $\boldsymbol{X}$ notation may imply that each element of a row can be individually selected, which is frequently not true.

Let m be the number of *mathematically independent* explanatory variables used in the linear combination of parameters, η. Denote these variables by $x_1, \ldots, x_m$. By "mathematically independent," I mean that these variables can take values independently of the values taken by other variables. By contrast, there may be variables used in η whose values are automatically fixed once the values of $x_1, \ldots, x_m$ are fixed. For example, if $\eta = \beta_0 + \beta_1 x_1 + \beta_2 x_2 + \beta_3 x_1 x_2$, there are $m = 2$ mathematically independent variables (x_1 and x_2). We cannot vary the value of $x_1 x_2$ once the values of x_1 and x_2 have been chosen. Denote by $\boldsymbol{x} = (x_1, \ldots, x_m)^\top$ a vector of the m explanatory variables.

Write $\boldsymbol{\beta} = (\beta_0, \ldots, \beta_{p-1})^\top$ for the $p \times 1$ vector containing the model parameters in the linear combination η. In η, the coefficient of each β_i will be a function of one or more of $x_1, \ldots, x_m$. We will write $f_i(\boldsymbol{x})$ for the function of $\boldsymbol{x}$ that is the coefficient of β_i ($i = 0, \ldots, p-1$), and denote by $\boldsymbol{f}(\boldsymbol{x})$ the vector of these functions. That is,

$$\boldsymbol{f}(\boldsymbol{x}) = \left(f_0(\boldsymbol{x}),\, f_1(\boldsymbol{x}), \ldots, f_{p-1}(\boldsymbol{x})\right)^\top.$$

These coefficients will be described as *regressors* in order to distinguish them from the individual explanatory variables.

Example 1.4.4. *Consider again the model $\eta = \beta_0 + \beta_1 x_1 + \beta_2 x_2 + \beta_3 x_1 x_2$. There are $m = 2$ explanatory variables (x_1 and x_2), $p = 4$ parameters ($\beta_0, \ldots, \beta_3$) and $p = 4$ regressors (1, x_1, x_2 and $x_1 x_2$). We* *have* $\boldsymbol{x} = (x_1, x_2)^\top$, $\boldsymbol{f}(\boldsymbol{x}) = (1, x_1, x_2, x_1x_2)^\top$, $\boldsymbol{\beta} = (\beta_0, \ldots, \beta_3)^\top$ *and* $\eta = \boldsymbol{f}^\top(\boldsymbol{x})\boldsymbol{\beta}$.

Let there be s *distinct values of the vector* $\boldsymbol{x}$ *in the experiment. These will be denoted by* $\boldsymbol{x}_1, \ldots, \boldsymbol{x}_s$, *and* $\boldsymbol{x}_i$ *will be called the* i*th* support point *of the design. Denote by* $\boldsymbol{F}$ *the* $s \times p$ *matrix whose* i*th row is* $\boldsymbol{f}^\top(\boldsymbol{x}_i)$ *(sometimes abbreviated to* $\boldsymbol{f}_i^\top$*). That is,*

$$\boldsymbol{F} = \begin{bmatrix} \boldsymbol{f}^\top(\boldsymbol{x}_1) \\ \vdots \\ \boldsymbol{f}^\top(\boldsymbol{x}_s) \end{bmatrix} = \begin{bmatrix} \boldsymbol{f}_1^\top \\ \vdots \\ \boldsymbol{f}_s^\top \end{bmatrix}.$$

Note that the matrices $\boldsymbol{F}$ *and* $\boldsymbol{X}$ *are equal only if there is just one observation made at each support point (and so* $s = N$*). If the* i*th support point* $\boldsymbol{x}_i$ *has* n_i *observations made on it* $(i = 1, \ldots, s)$, *then* $\boldsymbol{f}^\top(\boldsymbol{x}_i)$ *will occur in* n_i *rows of* $\boldsymbol{X}$. *It follows that*

$$\boldsymbol{X}^\top\boldsymbol{X} = \boldsymbol{F}^\top \mathit{diag}(n_1, \ldots, n_s)\boldsymbol{F}.$$

Example 1.4.5. *Suppose that a design has* $s = 3$ *support points* $\boldsymbol{x}_1$, $\boldsymbol{x}_2$ *and* $\boldsymbol{x}_3$, *and independent observations are made on these points* $n_1 = 3$, $n_2 = 2$ *and* $n_3 = 1$ *times, respectively. Then*

$$\boldsymbol{X} = \begin{bmatrix} \boldsymbol{f}_1^\top \\ \boldsymbol{f}_1^\top \\ \boldsymbol{f}_1^\top \\ \boldsymbol{f}_2^\top \\ \boldsymbol{f}_2^\top \\ \boldsymbol{f}_3^\top \end{bmatrix},$$

and

$$\begin{aligned} \boldsymbol{X}^\top\boldsymbol{X} &= \boldsymbol{f}_1\boldsymbol{f}_1^\top + \boldsymbol{f}_1\boldsymbol{f}_1^\top + \boldsymbol{f}_1\boldsymbol{f}_1^\top + \boldsymbol{f}_2\boldsymbol{f}_2^\top + \boldsymbol{f}_2\boldsymbol{f}_2^\top + \boldsymbol{f}_3\boldsymbol{f}_3^\top \\ &= 3\boldsymbol{f}_1\boldsymbol{f}_1^\top + 2\boldsymbol{f}_2\boldsymbol{f}_2^\top + \boldsymbol{f}_3\boldsymbol{f}_3^\top \\ &= [\boldsymbol{f}_1, \boldsymbol{f}_2, \boldsymbol{f}_3]\ \mathit{diag}(3,2,1) \begin{bmatrix} \boldsymbol{f}_1^\top \\ \boldsymbol{f}_2^\top \\ \boldsymbol{f}_3^\top \end{bmatrix} \\ &= \boldsymbol{F}^\top \mathit{diag}(3,2,1)\boldsymbol{F}. \end{aligned}$$

1.5 Estimating the values of the model parameters

Two common methods of estimating $\boldsymbol{\beta}$ in general linear models are least squares (LS) estimation and maximum likelihood (ML) estimation. In

LS estimation, the estimate of $\boldsymbol{\beta}$ is that value of $\boldsymbol{\beta}$ that minimises the sum of squares of the error terms in the model (1.6). That is, the LS estimator is the value of $\boldsymbol{\beta}$ that minimises

$$\sum_{i=1}^{N} E_i^2 = \boldsymbol{E}^\top \boldsymbol{E} = (\boldsymbol{Y} - \boldsymbol{X\beta})^\top (\boldsymbol{Y} - \boldsymbol{X\beta}).$$

However, the assumptions underlying this estimation process are not generally applicable for distributions from the exponential family other than the normal, so LS estimation is not usually considered in GLMs.

It is customary to consider the ML estimators of $\beta_0, \ldots, \beta_{p-1}$. The vector of ML estimators, $\hat{\boldsymbol{\beta}} = (\hat{\beta}_0, \ldots, \hat{\beta}_{p-1})^\top$, consists of those values of $\beta_0, \ldots, \beta_{p-1}$ that maximise the likelihood of an observed random sample $y_1, \ldots, y_N$ from the distribution.

The *likelihood* is regarded as a function of the elements of $\boldsymbol{\beta}$. When the observations are independent, the likelihood is defined by

$$L(\boldsymbol{\beta}; y_1, \ldots, y_N) = f_Y(y_1; \theta_1) \times \ldots \times f_Y(y_N; \theta_N) = \prod_{i=1}^{N} f_Y(y_i; \theta_i). \quad (1.16)$$

As $\ln L(\boldsymbol{\beta}; y_1, \ldots, y_N)$ achieves its maximum at the same value of $\boldsymbol{\beta}$ as does $L(\boldsymbol{\beta}; y_1, \ldots, y_N)$, and as it is usually easier to find derivatives of $\ln L$ than of L, it is customary to seek that value of $\boldsymbol{\beta}$ for which $\ln L(\boldsymbol{\beta})$ is maximised. It is usual to denote $\ln L(\boldsymbol{\beta}; y_1, \ldots, y_N)$ by $\ell(\boldsymbol{\beta}; y_1, \ldots, y_N)$.

The quantity $\ell(\boldsymbol{\beta})$ will be maximised when its partial derivatives satisfy

$$\left. \frac{\partial \ell}{\partial \beta_i} \right|_{\boldsymbol{\beta} = \hat{\boldsymbol{\beta}}} = 0 \quad (i = 0, 1, \ldots, p-1), \quad (1.17)$$

and when the matrix of partial second derivatives, whose (j, k) element is $\partial^2 \ell / (\partial \beta_j \partial \beta_k)$, is *negative definite* (defined on page 23). Equation (1.17) can also be written in vector form as

$$\left. \frac{\partial \ell}{\partial \boldsymbol{\beta}} \right|_{\boldsymbol{\beta} = \hat{\boldsymbol{\beta}}} = \mathbf{0}. \quad (1.18)$$

It is customary to write

$$U_j = \frac{\partial \ell}{\partial \beta_j},$$

where U_j is called the jth *score statistic*. Define the $p \times 1$ vector $\boldsymbol{U}$ by $\boldsymbol{U} = (U_0, \ldots, U_{p-1})^\top$.

For distributions from the exponential family that have the canonical form, it can be shown (e.g., Dobson & Barnett, 2008, Eq. (4.18)) that

$$U_j = \sum_{i=1}^{N} \left[\frac{(Y_i - \mu_i)}{\text{var}(Y_i)} x_{ij} \left(\frac{\partial \mu_i}{\partial \eta_i} \right) \right], \quad j \in \{0, \ldots, p-1\}, \tag{1.19}$$

where x_{ij} is the (i, j) element of the design matrix $\boldsymbol{X}$. If the design matrix is represented by $\boldsymbol{F}$, where each row of $\boldsymbol{F}$ is unique,

$$U_j = \sum_{i=1}^{s} n_i \left[\frac{(Y_i - \mu_i)}{\text{var}(Y_i)} f_{ij} \left(\frac{\partial \mu_i}{\partial \eta_i} \right) \right], \quad j \in \{0, \ldots, p-1\}, \tag{1.20}$$

where f_{ij} is the (i, j) element of $\boldsymbol{F}$. Then $\text{E}(U_j) = 0$ for each j, which implies that $\text{E}(\boldsymbol{U}) = \boldsymbol{0}$.

In addition, denote by $\boldsymbol{\mathcal{I}}$ the $p \times p$ covariance matrix of $\boldsymbol{U}$. Adapting (4.20) of Dobson & Barnett (2008), the (j, k) element of $\boldsymbol{\mathcal{I}}$, $\mathcal{I}_{jk} = \text{cov}(U_j, U_k)$, is given by

$$\mathcal{I}_{jk} = \sum_{i=1}^{s} n_i \frac{f_{ij} f_{ik}}{\text{var}(Y_i)} \left(\frac{\partial \mu_i}{\partial \eta_i} \right)^2, \quad j, k \in \{0, \ldots, p-1\}. \tag{1.21}$$

These results are important in calculating the ML estimate $\hat{\boldsymbol{\beta}}$. However, of most relevance to us when considering the design of experiments that will involve GLMs are the results that, asymptotically (i.e., for very large values of N),

$$\text{E}(\hat{\boldsymbol{\beta}}) = \boldsymbol{\beta} \quad \text{and} \quad \text{cov}(\hat{\boldsymbol{\beta}}) = \boldsymbol{\mathcal{I}}^{-1}. \tag{1.22}$$

For an experiment involving normally distributed observations with a constant variance, the results in (1.22) are exact for all values of N, and are equivalent to those in (1.8) and (1.9). For any non-normal distribution, although the equations in (1.22) are asymptotic results only, they are generally used for all values of N, because it is difficult to obtain exact expressions for $\text{E}(\hat{\boldsymbol{\beta}})$ and $\text{cov}(\hat{\boldsymbol{\beta}})$ for small values of N.

Example 1.5.1. *Consider a standard straight line regression model $Y_i = \beta_0 + \beta_1 x_i + E_i$ $(i = 1, \ldots, n)$. The linear combination of model parameters can be written as $\eta = \boldsymbol{f}^\top(x)\boldsymbol{\beta}$, where $\boldsymbol{f}^\top(x) = (1, x)$ and $\boldsymbol{\beta} = (\beta_0, \beta_1)^\top$. Let the set of possible values of x be $\mathcal{X} = \{x : 0 \leq x \leq 1\}$. Consider two possible experimental designs, each with $n = 10$ observations. Design 1, the "equispaced" design, places one observation at each of $0, 1/9, 2/9, \ldots, 1$. Design 2, the "extremities" design, places five observations at each of 0 and 1. Then the design matrices are*

Design 1

$$\boldsymbol{X}_1^\top = \begin{bmatrix} 1 & 1 & 1 & 1 & 1 & 1 & 1 & 1 & 1 & 1 \\ 0 & \frac{1}{9} & \frac{2}{9} & \frac{3}{9} & \frac{4}{9} & \frac{5}{9} & \frac{6}{9} & \frac{7}{9} & \frac{8}{9} & 1 \end{bmatrix}$$

$$\boldsymbol{F}_1^\top = \begin{bmatrix} 1 & 1 & 1 & 1 & 1 & 1 & 1 & 1 & 1 & 1 \\ 0 & \frac{1}{9} & \frac{2}{9} & \frac{3}{9} & \frac{4}{9} & \frac{5}{9} & \frac{6}{9} & \frac{7}{9} & \frac{8}{9} & 1 \end{bmatrix}^\top;$$

$$n_1 = \ldots = n_{10} = 1;$$

$$\boldsymbol{X}_1^\top \boldsymbol{X}_1 = \boldsymbol{F}_1^\top \, diag(1, 1, \ldots, 1) \boldsymbol{F}_1 = \frac{1}{81} \begin{bmatrix} 810 & 405 \\ 405 & 285 \end{bmatrix}.$$

Design 2

$$\boldsymbol{X}_2^\top = \begin{bmatrix} 1 & 1 & 1 & 1 & 1 & 1 & 1 & 1 & 1 & 1 \\ 0 & 0 & 0 & 0 & 0 & 1 & 1 & 1 & 1 & 1 \end{bmatrix}$$

$$\boldsymbol{F}_2 = \begin{bmatrix} 1 & 0 \\ 1 & 1 \end{bmatrix};$$

$$n_1 = n_2 = 5;$$

$$\boldsymbol{X}_2^\top \boldsymbol{X}_2 = \boldsymbol{F}_2^\top \, diag(5, 5) \boldsymbol{F}_2 = \begin{bmatrix} 10 & 5 \\ 5 & 5 \end{bmatrix}.$$

Using the result in (1.9) *for* $cov(\hat{\boldsymbol{\beta}})$, *it follows that, for Design 1,*

$$cov(\hat{\boldsymbol{\beta}}) = \sigma^2 (\boldsymbol{X}_1^\top \boldsymbol{X}_1)^{-1} = \frac{\sigma^2}{55} \begin{bmatrix} 19 & -27 \\ -27 & 54 \end{bmatrix},$$

while, for Design 2,

$$cov(\hat{\boldsymbol{\beta}}) = \sigma^2 (\boldsymbol{X}_2^\top \boldsymbol{X}_2)^{-1} = \frac{\sigma^2}{5} \begin{bmatrix} 1 & -1 \\ -1 & 2 \end{bmatrix}.$$

For the straight line regression model, $g(\mu) = \mu = \eta = \beta_0 + \beta_1 x = \boldsymbol{f}^\top(x)\boldsymbol{\beta}$, *where* $\boldsymbol{f}^\top = (1, x)$. *Then*

$$\frac{\partial \mu_i}{\partial \eta_i} = 1 \quad and \quad f_{i0} = 1, \; f_{i1} = x_i \quad (i = 1, \ldots, s).$$

Recall from (1.21) *that the elements of the covariance matrix* $\boldsymbol{\mathcal{I}}$ *are*

$$\mathcal{I}_{jk} = \sum_{i=1}^{s} n_i \frac{f_{ij} f_{ik}}{var(Y_i)} \left(\frac{\partial \mu_i}{\partial \eta_i} \right)^2, \quad j, k \in \{0, \ldots, p-1\}.$$

As $var(Y_i) = \sigma^2$ *for each* $i = 1, \ldots, s$, *then it follows that*

$$\mathcal{I}_{jk} = \frac{1}{\sigma^2} \sum_{i=1}^{s} n_i f_{ij} f_{ik} \left(\frac{\partial \mu_i}{\partial \eta_i} \right)^2, \quad j, k \in \{0, \ldots, p-1\}.$$

For the straight line regression model, the elements of the matrix $\mathcal{I}$ *are*

$$\begin{aligned}
\mathcal{I}_{00} &= \frac{1}{\sigma^2}\sum_{i=1}^{s} n_i(1\times 1)(1)^2,\\
\mathcal{I}_{01} = \mathcal{I}_{10} &= \frac{1}{\sigma^2}\sum_{i=1}^{s} n_i(1\times x_i)(1)^2,\\
\mathcal{I}_{11} &= \frac{1}{\sigma^2}\sum_{i=1}^{s} n_i(x_i\times x_i)(1)^2,
\end{aligned}$$

and thus

$$\mathcal{I} = \frac{1}{\sigma^2}\begin{bmatrix} \sum_{i=1}^{s} n_i & \sum_{i=1}^{s} n_i x_i \\ \sum_{i=1}^{s} n_i x_i & \sum_{i=1}^{s} n_i x_i^2 \end{bmatrix}.$$

As $cov(\hat{\boldsymbol{\beta}}) = \mathcal{I}^{-1}$*, this gives the same expression for* $cov(\hat{\boldsymbol{\beta}})$ *as from* $cov(\hat{\boldsymbol{\beta}}) = \sigma^2(\mathbf{X}^\top\mathbf{X})^{-1}$ *in* (1.10).

So it has been demonstrated, for the straight line regression model and a normal distribution with a constant variance, that the asymptotic result for $\text{cov}(\hat{\boldsymbol{\beta}})$ from (1.22) is actually an exact result for any value of N, whether large or small. In fact, it is an exact result for any model that assumes that the observations are independent and from a normal distribution with a constant variance. For any other assumption or distribution, the result is only correct asymptotically.

Chapter 2

Background Material

2.1 Introduction

While the mathematical theory required in this book has been minimised as much as possible, one cannot follow the design theory without a minimal knowledge of calculus and matrix algebra. This chapter includes a brief overview of the fundamental points needed for the design of experiments. All calculations are done using a computing package. It is advantageous to choose one with statistical capabilities.

There are many statistical software packages available to perform data analysis. Look at the Wikipedia article "List of Statistical Packages" (Wikimedia, 2018) for a lengthy listing. Be sure that you scroll all the way to the bottom of the list; it requires more than one screen to see all the packages. Some of them have specialised uses, while others will perform many different types of analyses. Some are free, and several are expensive to purchase.

Some of these packages have features that assist you to design an experiment, but usually only for general linear models with most or all of the predictor variables being categorical in nature. A subset of packages has facilities that can be used to design experiments for GLMs with continuous variables, which are the experiments that we will mostly consider in this book. I am unaware of any package that has built-in facilities to design experiments involving GLMs. So a package is required that is flexible, and where one can write one's own commands (as opposed to being forced to use only the commands provided by the software writers), in order to produce the designs that will be needed.

The package that will be used in this book is called R (R Core Team, 2018). Its greatest attraction to many people is that it is completely free. However, this ignores many other favourable features of R. It has a very large community of users, some of whom are writing additional commands, and the versatility of the package continues to grow. It can be programmed by any user and so, if there is a feature that you want but cannot find, it is possible to write some R code yourself.

The package can be downloaded from *www.cran.r-project.org* in Windows, Mac OS or Linux versions. Many books on statistical analyses using R include some instructions on running the program, and there are also books available that are solely devoted to using R. The book by Crawley (2013) provides a vast array of information. Another useful book is de Micheaux, Drouilhet, & Liquet (2013). Many other books deal with specialised aspects of R. Help is readily available from online users groups and the Worldwide Web.

The present book does not aim to train you in the use of R. It assumes that you have some familiarity with the program, and concentrates on those features of R that are of specific use to design an experiment for a GLM. Each possible design will be characterised by a single number that is a function of various aspects of the design. The most important uses that we will have for R are

1. the maximisation or minimisation of this individual number;
2. simulation of data, either
   (a) as a way of obtaining an idea of the characteristics of the design for which a maximum or minimum occurs, or
   (b) in examining the properties of some allegedly optimal design;
3. analysing some data using a GLM, to help with point 2(b) above.

2.2 Maximisation or minimisation of a function

2.2.1 *A function of one variable*

Let us briefly revise the notion of finding the maximum or minimum of a function of one variable; e.g., $y = f(x) = x^2 - 2x - 3$. This requires a knowledge of elementary calculus. A *turning point* will occur at values of x for which $(dy)/(dx) = f'(x)$ is equal to 0. Suppose that this occurs at $x = a$. If $(d^2y)/(dx^2) < 0$ at $x = a$, the turning point is a maximum. If $(d^2y)/(dx^2) > 0$ at $x = a$, the turning point is a minimum. If $(d^2y)/(dx^2) = 0$ at $x = a$, the turning point is a point of inflection. Points of inflection will not be of interest in our work.

Example 2.2.1. *Let $y = f(x) = x^2 - 2x - 3$. Then $(dy)/(dx) = 2x - 2$ and $(d^2y)/(dx^2) = 2$. Clearly $(dy)/(dx) = 0$ when $x = 1$. Moreover, $(d^2y)/(dx^2) > 0$ for all values of x, so the turning point at $x = 1$ is a minimum. A maximum for $y = f(x)$ does not exist if no restriction is placed on the value of x, as is evident from the graph of $y = x^2 - 2x - 3$ in Figure 2.1.*

2.2.2 *A function of more than one variable*

Now consider the case where $\boldsymbol{x}$ is a vector of q mathematically independent variables, for some $q > 1$. Let y be a function of the variables in $\boldsymbol{x}$,

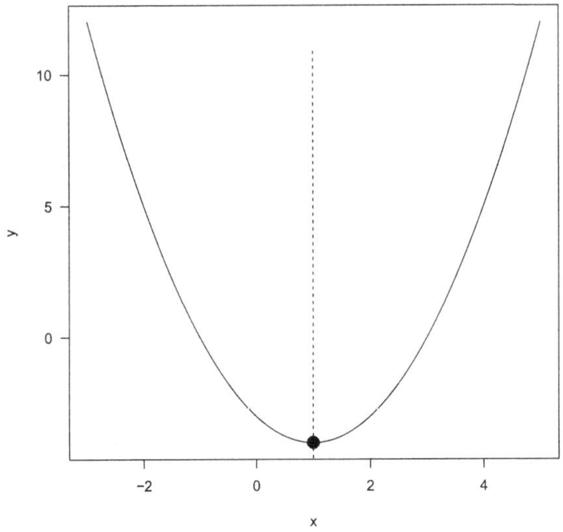

Figure 2.1 *A graph of $y = x^2 - 2x - 3$, showing that a minimum occurs at $x = 1$. The marked point is at $x = 1$, $y = -4$.*

$y = f(\boldsymbol{x})$. A $q \times 1$ vector $\boldsymbol{a}$ is a turning point of y if each of the q partial derivatives $(\partial y)/(\partial x_i)$ is equal to 0 at $\boldsymbol{x} = \boldsymbol{a}$. To determine whether $\boldsymbol{a}$ is a maximum, a minimum, or some other kind of turning point (e.g., a saddle point), one must calculate the matrix of partial second derivatives, $\boldsymbol{\mathcal{S}}$, whose (i, j) element is $(\partial^2 y)/(\partial x_i \partial x_j)$ $(i, j = 1, \ldots, q)$ and evaluate it at $\boldsymbol{x} = \boldsymbol{a}$. Call this matrix $\boldsymbol{\mathcal{S}}(\boldsymbol{a})$. The turning point will be a minimum if $\boldsymbol{\mathcal{S}}(\boldsymbol{a})$ is a *positive definite* matrix, and a maximum if $\boldsymbol{\mathcal{S}}(\boldsymbol{a})$ is a *negative definite* matrix. If $\boldsymbol{\mathcal{S}}(\boldsymbol{a})$ is neither negative definite nor positive definite, the turning point is neither a maximum nor a minimum.

The $q \times q$ matrix $\boldsymbol{\mathcal{S}}(\boldsymbol{a})$ is positive definite if, for any $q \times 1$ vector $\boldsymbol{z}$, the quantity $\boldsymbol{z}^\top \boldsymbol{\mathcal{S}}(\boldsymbol{a})\boldsymbol{z}$ is greater than 0 for every $\boldsymbol{z} \neq \mathbf{0}$. Similarly, $\boldsymbol{\mathcal{S}}(\boldsymbol{a})$ is negative definite if, for any $q \times 1$ vector $\boldsymbol{z}$, the quantity $\boldsymbol{z}^\top \boldsymbol{\mathcal{S}}(\boldsymbol{a})\boldsymbol{z}$ is less than 0 for every $\boldsymbol{z} \neq \mathbf{0}$.

Example 2.2.2. *Let y be a function of $q = 2$ mathematically independent variables x_1 and x_2. Write $\boldsymbol{x} = (x_1, x_2)^\top$ and let $y = f(\boldsymbol{x}) = x_1^2 + 4x_1x_2 + 6x_2^2 + 4x_1 + 4x_2 + 3$. Hence $(\partial y)/(\partial x_1) = 2x_1 + 4x_2 + 4$ and $(\partial y)/(\partial x_2) = 4x_1 + 12x_2 + 4$, so $(\partial y)/(\partial x_1) = 0$ and $(\partial y)/(\partial x_2) = 0$*

imply

$$\begin{aligned} 2x_1 + 4x_2 + 4 &= 0 \\ 4x_1 + 12x_2 + 4 &= 0. \end{aligned}$$

Solving these equations simultaneously gives $x_1 = -4$ *and* $x_2 = 1$*, so the function* $y = f(\boldsymbol{x})$ *has a turning point at* $\boldsymbol{a} = (-4, 1)^\top$*. Additionally,*

$$\frac{\partial^2 y}{\partial x_1^2} = 2, \; \frac{\partial^2 y}{\partial x_1 \partial x_2} = \frac{\partial^2 y}{\partial x_2 \partial x_1} = 4, \; \textit{and} \; \frac{\partial^2 y}{\partial x_2^2} = 12,$$

irrespective of the value of $\boldsymbol{a}$*, so*

$$\boldsymbol{S}(\boldsymbol{a}) = \begin{bmatrix} 2 & 4 \\ 4 & 12 \end{bmatrix}.$$

For $\boldsymbol{z} = (z_1, z_2)^\top$, $\boldsymbol{z}^\top \boldsymbol{S}(\boldsymbol{a})\boldsymbol{z} = 2z_1^2 + 8z_1z_2 + 12z_2^2 = 2(z_1^2 + 4z_1z_2 + 4z_2^2) + 4z_2^2 = 2(z_1 + 2z_2)^2 + 4z_2^2$*. This quantity can be zero only if* $z_1 + 2z_2 = 0$ *and* $z_2 = 0$ *simultaneously (i.e., if* $z_1 = z_2 = 0$*, or* $\boldsymbol{z} = \boldsymbol{0}$*); otherwise,* $\boldsymbol{z}^\top \boldsymbol{S}(\boldsymbol{a})\boldsymbol{z}$ *is greater than zero. Hence* $\boldsymbol{S}(\boldsymbol{a})$ *is positive definite, and so the turning point* $\boldsymbol{a} = (-4, 1)^\top$ *minimises* $y = f(\boldsymbol{x})$*.*

In Example 2.2.2, it was easy to demonstrate that $\boldsymbol{S}(\boldsymbol{a})$ is positive definite. However, as (i) the number of variables increases, or (ii) the complexity of the function $f(\boldsymbol{x})$ increases, a simple determination in this manner of whether $\boldsymbol{S}(\boldsymbol{a})$ is positive or negative definite becomes very much more difficult. It is much easier to use the following:

Result 2.2.1. *A* $q \times q$ *matrix* $\boldsymbol{A}$ *is positive (negative) definite if and only if each of its* q *eigenvalues is positive (negative).*

A detailed discussion of eigenvalues is beyond the scope of this book. The interested reader should consult an introductory text on linear algebra. Harville (1997) and Searle (1982) provide more detailed and advanced discussion. In brief, a $q \times 1$ vector $\boldsymbol{u}$ satisfying $\boldsymbol{u} \neq \boldsymbol{0}$ is called an *eigenvector* (sometimes "characteristic vector") of the $q \times q$ matrix $\boldsymbol{A}$ if $\boldsymbol{A}\boldsymbol{u} = \lambda \boldsymbol{u}$ for some λ; that is, the result of post-multiplying $\boldsymbol{A}$ by the vector $\boldsymbol{u}$ is a multiple of $\boldsymbol{u}$. The multiple, λ, is called the *eigenvalue* ("characteristic value") of $\boldsymbol{A}$ corresponding to the eigenvector $\boldsymbol{u}$.

Example 2.2.3. *Let* $q = 2$*. Consider the* $q \times q$ *matrix*

$$\boldsymbol{A} = \begin{bmatrix} 2 & 1 \\ 1 & 2 \end{bmatrix}. \tag{2.1}$$

The vector $\boldsymbol{u}_1 = (1,1)^\top$ *is an eigenvector of* $\boldsymbol{A}$ *with corresponding eigenvalue* $\lambda_1 = 3$*, as* $\boldsymbol{A}\boldsymbol{u}_1 = \boldsymbol{A}(1,1)^\top = (3,3)^\top = 3\boldsymbol{u}_1 = \lambda_1\boldsymbol{u}_1$*. The vector* $\boldsymbol{u}_2 = (1,-1)^\top$ *is also an eigenvector of* $\boldsymbol{A}$*, with corresponding eigenvalue* $\lambda_2 = 1$*, because* $\boldsymbol{A}\boldsymbol{u}_2 = \boldsymbol{A}(1,-1)^\top = (1,-1)^\top = 1\boldsymbol{u}_2 = \lambda_2\boldsymbol{u}_2$*. As both eigenvalues of the* 2×2 *matrix are positive, then* $\boldsymbol{A}$ *is positive definite.*

If $\boldsymbol{u}$ is an eigenvector of $\boldsymbol{A}$ with corresponding eigenvalue λ, then any nonzero multiple of $\boldsymbol{u}$, $k\boldsymbol{u}$ ($k \neq 0$), is also an eigenvector of $\boldsymbol{A}$ with corresponding eigenvalue λ. This follows since $\boldsymbol{A}(k\boldsymbol{u}) = k(\boldsymbol{A}\boldsymbol{u}) = k(\lambda\boldsymbol{u}) = \lambda(k\boldsymbol{u})$. To ensure that we are not regarding two vectors as different eigenvectors of $\boldsymbol{A}$ when they are really just multiples of one another, we can require an eigenvector to have the properties that

(a) it is *normalized* (i.e., its length, $\ell = \sqrt{(\boldsymbol{u}^\top\boldsymbol{u})}$, is equal to 1, which is achieved by using the particular multiple $(1/\ell)\boldsymbol{u}$), and

(b) its first nonzero element is positive.

Property (b) is not necessarily observed by computer packages, including R, so an eigenvector given by R might possibly be the negative of the one obtained by this rule. In Example 2.2.3, $\ell_1 = \sqrt{(\boldsymbol{u}_1^\top\boldsymbol{u}_1)} = \sqrt{2} = \sqrt{(\boldsymbol{u}_2^\top\boldsymbol{u}_2)} = \ell_2$. So $\boldsymbol{u}_1 = (1/\sqrt{2}, 1/\sqrt{2})^\top$ and $\boldsymbol{u}_2 = (1/\sqrt{2}, -1/\sqrt{2})^\top$ would be used as the eigenvectors of $\boldsymbol{A}$.

The calculation by hand of the eigenvalues and eigenvectors may sometimes be done for 2×2 and 3×3 matrices, but it is generally much more difficult for larger matrices. It is better to use a computer. R will calculate eigenvalues and eigenvectors, using the function *eigen*.

The following program

```
amatrix <- matrix(c(2,1,1,2),2,2)
out <- eigen(amatrix)
out$vectors
out$values
```

defines the matrix $\boldsymbol{A}$ and calculates the eigenvectors and corresponding eigenvalues of $\boldsymbol{A}$, then lists them. The output is as follows:

```
> out$vectors
          [,1]       [,2]
[1,] 0.7071068 -0.7071068
[2,] 0.7071068  0.7071068
> out$values
[1] 3 1
```

Each column in `out$vectors` represents an eigenvector. The eigenvalues match those specified earlier, and the first eigenvector is equal to $\boldsymbol{u}_1$. The second eigenvector is equal to $-\boldsymbol{u}_2$, so property (b) of eigenvectors is not possessed. This discrepancy is not of any importance, as our interest will lie in the eigenvalues, rather than the eigenvectors, of a matrix.

As R calculates the eigenvalues and eigenvectors *numerically,* the results may differ very slightly from those that would be obtained theoretically. In particular, if an eigenvalue differs from 0 by a very small amount (less than 10^{-6}, say), you should probably regard the true value of the eigenvalue as being 0.

While eigenvalues have been introduced as a means of determining whether a matrix is positive definite, negative definite, or neither of these, we will see in Chapter 3 other important uses for eigenvalues.

Several other important considerations involving matrices and their eigenvalues will be quickly described.

1. In the work here, we will always be concerned with symmetric matrices. A matrix $\boldsymbol{A}$ is symmetric if $\boldsymbol{A}^\top = \boldsymbol{A}$. For a symmetric $q \times q$ matrix $\boldsymbol{A}$, there are q real eigenvalues $\lambda_1, \ldots, \lambda_q$ (not necessarily distinct).
2. One number arising from $\boldsymbol{A}$ is called the *determinant* of $\boldsymbol{A}$, and is denoted by $\det(\boldsymbol{A})$ or $|\boldsymbol{A}|$. Its original definition is beyond the scope of this book, but see Harville (1997, Chapter 13) or Searle (1982, Chapter 4) for further information. However, we note here that $\det(\boldsymbol{A})$ is equal to the product of the eigenvalues of $\boldsymbol{A}$; i.e.,

$$\det(\boldsymbol{A}) = \prod_{i=1}^{q} \lambda_i.$$

   The matrix $\boldsymbol{A}$ in (2.1) has eigenvalues 3 and 1, so $\det(\boldsymbol{A}) = 3 \times 1 = 3$.
3. If $\boldsymbol{A} = \text{diag}(a_1, a_2, \ldots, a_q)$ is a *diagonal matrix* (all its off-diagonal elements are zero), then $\det(\boldsymbol{A}) = a_1 \times a_2 \times \cdots \times a_q$; i.e.,

$$\det(\boldsymbol{A}) = \prod_{i=1}^{q} a_i.$$

4. The determinant of a 2×2 matrix can be calculated very simply without recourse to eigenvalues.

   Result 2.2.2. *A* 2×2 *matrix*

$$\boldsymbol{A} = \begin{bmatrix} a & b \\ c & d \end{bmatrix}$$

   has $\det(\boldsymbol{A}) = ad - bc$.

   Thus, for $\boldsymbol{A}$ in (2.1), $\det(\boldsymbol{A}) = 2 \times 2 - 1 \times 1 = 3$ (as shown above).
5. If $\boldsymbol{A}$ is a $q \times q$ matrix and k is a constant, then $\det(k\boldsymbol{A}) = k^q \det(\boldsymbol{A})$. For example, if $\boldsymbol{A}$ is 4×4, then $\det(3\boldsymbol{A}) = 3^4 \det(\boldsymbol{A}) = 81 \det(\boldsymbol{A})$.

6. If $\det(\boldsymbol{A}) \neq 0$, then there exists a matrix, denoted by $\boldsymbol{A}^{-1}$ and called the *inverse* of $\boldsymbol{A}$, which satisfies $\boldsymbol{A}\boldsymbol{A}^{-1} = \boldsymbol{A}^{-1}\boldsymbol{A} = \boldsymbol{I}_q$, where $\boldsymbol{I}_q$ is the $q \times q$ identity matrix. The vector $\boldsymbol{u}$ is an eigenvector of $\boldsymbol{A}$ with corresponding eigenvalue λ if and only if $\boldsymbol{u}$ is an eigenvector of $\boldsymbol{A}^{-1}$ with eigenvalue λ^{-1}. [You can see why we require $\det(\boldsymbol{A}) \neq 0$. If $\det(\boldsymbol{A}) = \prod \lambda_i = 0$, then at least one of the eigenvalues of $\boldsymbol{A}$ equals zero, and we cannot find the reciprocal of 0.] If $\det(\boldsymbol{A}) \neq 0$, so $\boldsymbol{A}^{-1}$ exists, the property $\det(\boldsymbol{A}^{-1}) = [\det(\boldsymbol{A})]^{-1}$ holds.
7. When $\boldsymbol{A}$ and $\boldsymbol{B}$ are both $q \times q$ matrices, the result $\det(\boldsymbol{A}\boldsymbol{B}) = \det(\boldsymbol{A}) \times \det(\boldsymbol{B})$ holds. Also, $\det(\boldsymbol{A}^\top) = \det(\boldsymbol{A})$.
8. If $\boldsymbol{A}$ is a $q \times q$ symmetric matrix of constants and $\boldsymbol{x}$ is a $q \times 1$ vector of variables, then the scalar $\boldsymbol{x}^\top\boldsymbol{A}\boldsymbol{x}$ is called a *quadratic form.* This name comes because $\boldsymbol{x}^\top\boldsymbol{A}\boldsymbol{x}$ is the sum of terms that are all of the form $a_{ij}x_ix_j$, where a_{ij} is the (i,j) element of $\boldsymbol{A}$ and x_i is the ith element of $\boldsymbol{x}$. A quadratic involves the product of a variable by another (not necessarily distinct) variable. For example, if $q = 3$, $\boldsymbol{x} = (x_1, x_2, x_3)^\top$ and
$$\boldsymbol{A} = \begin{bmatrix} 2 & 1 & -1 \\ 1 & 3 & -2 \\ -1 & -2 & 4 \end{bmatrix},$$
then
$$\begin{aligned} \boldsymbol{x}^\top\boldsymbol{A}\boldsymbol{x} &= 2x_1^2 + x_1x_2 - x_1x_3 + x_2x_1 + 3x_2^2 - 2x_2x_3 - x_3x_1 - 2x_3x_2 + 4x_3^2 \\ &= 2x_1^2 + 2x_1x_2 - 2x_1x_3 + 3x_2^2 - 4x_2x_3 + 4x_3^2. \end{aligned}$$

Quadratic forms will arise in the selection of optimal designs, and we will be interested in maximising or minimising terms of the form $\boldsymbol{x}^\top\boldsymbol{A}\boldsymbol{x}$, subject to the restriction that $\boldsymbol{x}^\top\boldsymbol{x} = 1$; i.e., that $\boldsymbol{x}$ is normalized. The following result is what is needed.

Result 2.2.3. *Let $\boldsymbol{A}$ be a $q \times q$ symmetric matrix of constants and $\boldsymbol{x}$ be a normalized $q \times 1$ vector of variables; i.e., $\boldsymbol{x}^\top\boldsymbol{x} = 1$. Denote by $\lambda_{\max}$ and $\lambda_{\min}$ the maximum and minimum eigenvalues of $\boldsymbol{A}$, respectively, and let $\boldsymbol{u}_{\max}$ and $\boldsymbol{u}_{\min}$ be the corresponding normalized eigenvectors. Then*

- *the maximum value of $\boldsymbol{x}^\top\boldsymbol{A}\boldsymbol{x}$ is $\lambda_{\max}$, and occurs when $\boldsymbol{x} = \boldsymbol{u}_{\max}$;*
- *the minimum value of $\boldsymbol{x}^\top\boldsymbol{A}\boldsymbol{x}$ is $\lambda_{\min}$, and occurs when $\boldsymbol{x} = \boldsymbol{u}_{\min}$.*

The proof of Result 2.2.3 is beyond the scope of this book, but it may be easily demonstrated for the 2×2 matrix $\boldsymbol{A}$ given in (2.1). It is known from the earlier examination of $\boldsymbol{A}$ that its two eigenvalues are $\lambda_{\max} = 3$ and $\lambda_{\min} = 1$, with corresponding normalized eigenvectors $\boldsymbol{u}_{\max} = (1/\sqrt{2}, 1/\sqrt{2})^\top$ and $\boldsymbol{u}_{\min} = (1/\sqrt{2}, -1/\sqrt{2})^\top$. Then Result 2.2.3 says that, for normalized $\boldsymbol{x}$, the maximum value of $\boldsymbol{x}^\top\boldsymbol{A}\boldsymbol{x}$ is 3, and occurs when

$\boldsymbol{x} = \boldsymbol{u}_{\max}$. The minimum value is 1, and occurs when $\boldsymbol{x} = \boldsymbol{u}_{\min}$. This will now be verified.

A normalized 2×1 vector may be written as $\boldsymbol{x} = (\sqrt{(1-z^2)}, z)^\top$, where $-1 \leq z \leq 1$. It is clear that $\boldsymbol{x}$ is normalized, as $\boldsymbol{x}^\top\boldsymbol{x} = (1-z^2) + z^2 = 1$. Under this definition, the first component of $\boldsymbol{x}$ is nonnegative, while the second component may be positive or negative. If you are concerned that this might be "cheating," please note that each component could be replaced by its negative without altering the result, as $(-\boldsymbol{x})^\top\boldsymbol{A}(-\boldsymbol{x}) = (-1)^2\boldsymbol{x}^\top\boldsymbol{A}\boldsymbol{x} = \boldsymbol{x}^\top\boldsymbol{A}\boldsymbol{x}$.

Then

$$\begin{aligned}
q &= \boldsymbol{x}^\top\boldsymbol{A}\boldsymbol{x} = \left[\sqrt{(1-z^2)}, z\right]\begin{bmatrix} 2 & 1 \\ 1 & 2 \end{bmatrix}\begin{bmatrix} \sqrt{(1-z^2)} \\ z \end{bmatrix} \\
&= 2(1-z^2) + 2\sqrt{(1-z^2)}z + 2z^2, \\
&= 2 + 2z\sqrt{(1-z^2)}, \text{ so} \\
\frac{dq}{dz} &= \frac{2-4z^2}{(1-z^2)^{1/2}}, \text{ and} \\
\frac{d^2q}{dz^2} &= \frac{-6z}{(1-z^2)^{3/2}}.
\end{aligned}$$

So $(dq)/(dz) = 0$ when $2 - 4z^2 = 0$, or $z = \pm 1/\sqrt{2}$. When $z = 1/\sqrt{2}$, $(d^2q)/(dz^2) < 0$, so the turning point is a maximum, and direct substitution shows that $q = 3 = \lambda_{\max}$ and $\boldsymbol{x} = (1/\sqrt{2}, 1/\sqrt{2})^\top = \boldsymbol{u}_{\max}$. When $z = -1/\sqrt{2}$, $(d^2q)/(dz^2) > 0$, so the turning point is a minimum; substitution of $z = -1/\sqrt{2}$ shows that $q = 1 = \lambda_{\min}$ and $\boldsymbol{x} = (1/\sqrt{2}, -1/\sqrt{2})^\top = \boldsymbol{u}_{\min}$. This confirms what was stated in Result 2.2.3.

Figure 2.2 shows a plot of $\boldsymbol{x}^\top\boldsymbol{A}\boldsymbol{x}$ vs. the value of z from which $\boldsymbol{x}$ is determined. The graph supports the conclusions drawn above for the values of the maximum and minimum of $\boldsymbol{x}^\top\boldsymbol{A}\boldsymbol{x}$ and the values of $\boldsymbol{x}$ that give these values.

2.3 Restrictions on independent variables

It was shown in Example 2.2.1 that $y = x^2 - 2x - 3$ has a minimum at $x = 1$, but that no maximum exists. This is obvious from Figure 2.1: $y \to \infty$ as $x \to \pm\infty$. However, if the values of x are restricted to the set $\mathcal{X}_1 = \{x : -1 \leq x \leq 3\}$, it is seen that y has two equal maxima of 0, at $x = -1$ and $x = 3$. If x is required to belong to $\mathcal{X}_2 = \{x : 0 \leq x \leq 4\}$, the global (or "overall") maximum is $y = 5$ at $x = 4$, and a local maximum is $y = 0$ at $x = 0$. None of the points mentioned here (which all lie on the boundaries of the sets of x-values) is indicated by standard calculus.

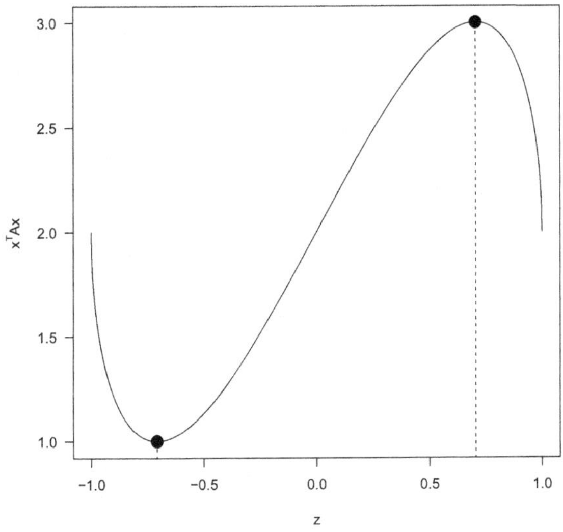

Figure 2.2 *A plot of* $q = \boldsymbol{x}^\top \boldsymbol{A}\boldsymbol{x}$ *vs* $z \in \{z : -1 \leq z \leq 1\}$*, where* $\boldsymbol{A}$ *is given in* (2.1) *and* $\boldsymbol{x} = (\sqrt{(1-z^2)}, z)^\top$*. The marked points are a minimum at* $(x, y) = (-1/\sqrt{2},\ 1)$ *and a maximum at* $(x, y) = (1/\sqrt{2},\ 4)$.

Denote by $\mathbb{R}$ the set of real numbers, and by $\mathbb{R}^q$ the set of $q \times 1$ vectors whose every element is a real number; $\mathbb{R}^q$ is often called "q-dimensional space." For a $q \times 1$ vector of mathematically independent variables, $\boldsymbol{x}$, locating a minimum or maximum value of $y = f(\boldsymbol{x})$ in a restricted segment of $\mathbb{R}^q$ is a nontrivial problem. Finding a maximum or minimum of a function is often called *optimising* the function. When this is done under some restrictions (or constraints) on the possible values that $\boldsymbol{x}$ can take, the process is called *constrained optimisation.*

The problem of finding an optimal design for some GLM will generally reduce to optimising some numerical property of the design with some constraints in place: constrained optimisation. A preliminary examination of how this is done in R will be made by examining some properties of potential designs considered earlier in Chapter 1.

Consider the straight line regression model $Y_i = \beta_0 + \beta_1 x + E_i$ $(i = 1, \ldots, N)$, where the E_i are considered to be independent observations from a $N(0, \sigma^2)$ distribution. From (1.9), (1.10) and the material immediately below these equations, the vector of ML estimates, $\hat{\boldsymbol{\beta}} = (\hat{\beta}_0, \hat{\beta}_1)^\top$, has covariance matrix

$$\frac{\sigma^2}{N\sum_{i=1}^N(x_i-\bar{x})^2}\begin{bmatrix}\sum_{i=1}^N x_i^2 & -\sum_{i=1}^N x_i\\ -\sum_{i=1}^N x_i & N\end{bmatrix}. \tag{2.2}$$

In Example 1.5.1, values of x were constrained to belong to the set $\mathcal{X} = \{x : 0 \leq x \leq 1\}$. Two experimental designs, each with $n = 10$ observations, were considered. Design 1, an "equispaced" design, had one observation at each of $0, 1/9, 2/9, \ldots, 1$, while Design 2, an "extremities" design, placed five observations at each of 0 and 1. Denote by $\boldsymbol{V}_1$ and $\boldsymbol{V}_2$ the values of $\text{cov}(\hat{\boldsymbol{\beta}})$ under Designs 1 and 2, respectively. It was shown in Example 1.5.1 that

$$\boldsymbol{V}_1^{-1} = \frac{\sigma^2}{55}\begin{bmatrix}19 & -27\\ -27 & 54\end{bmatrix} \quad \text{and} \quad \boldsymbol{V}_2^{-1} = \frac{\sigma^2}{5}\begin{bmatrix}1 & -1\\ -1 & 2\end{bmatrix}.$$

Suppose that we wish to choose between Designs 1 and 2 by selecting the one with the smaller value of $\text{var}(\hat{\beta}_1)$. Clearly Design 2 is to be preferred as it gives $\text{var}(\hat{\beta}_1) = 2\sigma^2/5 = 0.4\sigma^2$, while Design 1 gives $\text{var}(\hat{\beta}_1) = 54\sigma^2/55 = 0.9818\sigma^2$. To facilitate the comparison of the two designs, it is customary to assume that the value of σ^2 is the same for the two designs.

However, the aim in this book is to find optimal designs, which in this case would mean the design for 10 observations from $\mathcal{X}$ that has the *overall* minimum value of $\text{var}(\hat{\beta}_1)$. From (2.2), the general equation for $\text{var}(\hat{\beta}_1)$ when $N = 10$ is

$$\text{var}(\hat{\beta}_1) = \frac{\sigma^2}{10\sum_{i=1}^{10}(x_i-\bar{x})^2} \times 10 = \frac{\sigma^2}{\sum_{i=1}^{10}(x_i-\bar{x})^2}. \tag{2.3}$$

This expression is minimised by maximising $\sum_{i=1}^{10}(x_i-\bar{x})^2$. It is evident that this is done by maximising the distance of each x_i from $\bar{x}$, the average of all 10 x_i, and that, for $x_1, \ldots, x_{10} \in \mathcal{X}$, this is achieved by setting five of the x_i equal to 0 and the remaining five x_i equal to 1. In other words, Design 2 is the optimal design under the criterion of minimising $\text{var}(\hat{\beta}_1)$ for $n = 10$ observations from $\mathcal{X}$. No computing was required to find this optimum design.

Of course, the minimisation of $\text{var}(\hat{\beta}_1)$ is only one possible criterion for selecting a design. There are several others that might be considered. Suppose that we had wanted to minimise $\text{var}(\hat{\beta}_0)$ in (2.2).

This is being considered as an example only. It is unlikely that it would be regarded as more important to minimise $\text{var}(\hat{\beta}_0)$ than to minimise

$\text{var}(\hat{\beta}_1)$. Direct evaluation shows that $\text{var}(\hat{\beta}_0) = 19\sigma^2/55 \approx 0.34545\sigma^2$ for the "equispaced" design, whereas $\text{var}(\hat{\beta}_0) = 0.2\sigma^2$ for the "extremities" design. So clearly the latter design is also the better of the two designs for minimising $\text{var}(\hat{\beta}_0)$. *But is it the optimal design?*

Program_1 in the Web site doeforglm.com generates 10 values from the uniform distribution on the interval from $x = 0$ to $x = 1$, then calculates

$$V = \sum_{i=1}^{10} x_i^2 \Big/ [10 \sum_{i=1}^{10} (x_i - \bar{x})^2] = \sum_{i=1}^{10} x_i^2 \Big/ [10 \sum_{i=1}^{10} x_i^2 - (\sum_{i=1}^{10} x_i)^2] \quad (2.4)$$

The transition from the first expression for V to the second uses (1.11).

The generation of x-values and calculation of V is done repeatedly, the number of times being determined by the value of the program parameter *nsimulations*. Whenever a smaller value of V is detected, the new minimum and the corresponding set of 10 x-values are recorded.

A seed for the simulations can be specified. A *seed* is a numerical input to a random generation process that starts the process at a specific (but unknown) random number. The advantage of specifying a seed is that repeating the calculations using exactly the same data (e.g., if you modify the program and want to see if you get the same results as before) can be done by specifying the same seed again. If you have no interest in using the same randomly generated data a second time, the easiest thing to do is simply not to specify a seed, and R will generate one automatically without telling you what it is.

For *seed* = 1234 and *nsimulations* = 10000, a value of $V = 0.1621958$ was obtained. This shows that neither Design 1 nor Design 2 is optimal for minimising $\text{var}(\hat{\beta}_0)$. Increasing *nsimulations* and altering the seed gave further reductions in the value of V, until finally a value of $V = 0.123704$ was obtained from *seed* = 2245 and *nsimulations* = 20 million. After sorting from smallest to largest, and rounding to four decimal places, the 10 values of x were 0.0026, 0.0076, 0.0088, 0.0266, 0.0450, 0.0523, 0.0564, 0.0631, 0.1301, and 0.9665. It might not be possible to control x to four decimal places, so let us investigate the effect on the variance of rounding each x-value to two decimal places instead. The value of V is now 0.123958, a slight increase.

Performing twenty million simulations is not really sensible, for two reasons. One is that it is "over-kill," in the sense that 200 simulations would be adequate to ascertain that neither of the candidate designs is optimal, as other designs give smaller values of V. The second reason is that twenty million simulations is insufficient to find the actual optimal design; the best that it will do is to give a design that is approaching optimality.

Instead, we need to use a program that is designed to search for an optimum (either a minimum or a maximum, as appropriate; here it is a minimum). Applied mathematicians have devised many optimisation routines to locate an optimum. However, not all of them are useful in the present situation. A procedure is required that will minimise the variance in (2.4) subject to the constraints that $0 \le x_i \le 1$ for each $i = 1, \ldots, 10$. This is an application of *constrained optimisation.*

2.4 Constrained optimisation in R

2.4.1 The function constrOptim

An R function that will perform constrained optimisation is *constrOptim*. In this section, the input to *constrOptim* is denoted by a vector $\boldsymbol{v}$. Note that, here, $\boldsymbol{v}$ represents all possible variables that may be of interest, and not just the set of mathematically independent explanatory variables used to model the link function $g(\mu)$. The function *constrOptim* is a minimisation routine, but can be used for maximisation by changing the function to be optimised from $f(\boldsymbol{v})$ to $-f(\boldsymbol{v})$, as the value of $\boldsymbol{v}$ that minimises $f(\boldsymbol{v})$ will maximise $-f(\boldsymbol{v})$.

In its simplest form, *constrOptim* requires you to specify

1. the function of $\boldsymbol{v}$ to be minimised,
2. an initial guess of the value of $\boldsymbol{v}$ where the minimum occurs, and
3. the constraints that the variables in $\boldsymbol{v}$ must satisfy.

Consider these items in turn.

- The argument of $f(\boldsymbol{v})$ must be the full vector $\boldsymbol{v}$. You cannot input just some of the values of $\boldsymbol{v}$ unless you have modified the definition of $f(\boldsymbol{v})$.
- An initial guess at the value of $\boldsymbol{v}$ where $f(\boldsymbol{v})$ is minimised can be any vector within the region of interest, *provided that it does not lie on the boundary of that region.* The optimum may lie on the boundary, but the initial guess must not.
- Suppose that there are c constraints. These must be specified to *constrOptim* in the matrix form

  $$\boldsymbol{C}\boldsymbol{v} - \boldsymbol{u} \ge \boldsymbol{0}_c,$$

  where $\boldsymbol{v}$ is the $v \times 1$ vector of variables, $\boldsymbol{C}$ is a $c \times v$ matrix of constants, $\boldsymbol{u}$ is a $c \times 1$ vector of constants, and $\boldsymbol{0}_c$ is the $c \times 1$ vector whose every element is zero. The values of $\boldsymbol{C}$ and $\boldsymbol{u}$ form input to *constrOptim*.

While *constrOptim* can find minima that lie on the boundary of the region of interest, be aware that this may be a lengthy process. To quote from part of the information obtained by typing `help(constrOptim)` into R: *Minima in the interior of the feasible region are typically found quite quickly, but a substantial number of outer iterations may be needed for a minimum on the boundary.*

Example 2.4.1. *We return to the task of minimising var($\hat{\beta}_0$), and follow the three dot points above.*

- *As σ^2 is assumed to be constant for all designs, it can be removed from the calculations. So the function of $\boldsymbol{v} = (x_1, \ldots, x_{10})^\top$ to be optimised is*

$$
\begin{aligned}
h(\boldsymbol{v}) = \frac{1}{\sigma^2} var(\hat{\beta}_0) &= \frac{\sum_{i=1}^{10} x_i^2}{10\sum_{i=1}^{10}(x_i - \bar{x})^2} \\
&= \sum_{i=1}^{10} x_i^2 \Bigg/ \left[10\sum_{i=1}^{10} x_i^2 - (\sum_{i=1}^{10} x_i)^2 \right]. \quad (2.5)
\end{aligned}
$$

  The R function

```
varbeta0hat <- function(x)
{
 a <- sum(x^2)
 v <- a/(10*a - (sum(x))^2)
 v
}
```

  (which appears online in doeforglm.com as Program_2) takes a vector $\boldsymbol{x}$ of length 10 and calculates the required function. An R function that works for a vector of arbitrary length N would require only one more command, and appears in the online resources as Program_3, but I have chosen to illustrate the optimisation with a vector of specific length 10.

- *A vector of 10 values selected from the uniform distribution (e.g.,* `x <- runif(10)`*) would provide an acceptable starting point for constrOptim. However, you need try only a few starting points generated this way to discover that constrOptim may sometimes halt at a local minimum rather than at the global minimum.(This is like finding yourself at the bottom of a mine shaft, but not at the bottom of the deepest mine shaft). Using a good estimate of the value of $\boldsymbol{x}$ that gives an optimum will save lots of time.*

  In this case, my excessive zeal in running 20 million simulations has at least given a guess of the optimal design that should be "good": 0.0026, 0.0076, 0.0088, 0.0266, 0.0450, 0.0523, 0.0564, 0.0631, 0.1301, and 0.9665. None of these values lies on either boundary of the design space $\{x : 0 \leq x \leq 10\}$.

- *Consider the constraint* $0 \leq x_1 \leq 1$. *This can be broken into two parts: (i)* $0 \leq x_1$, *which is equivalent to* $1 \times x_1 - 0 \geq 0$, *and (ii)* $x_1 \leq 1$, *which is equivalent to* $-x_1 \geq -1$, *or* $(-1) \times x_1 - (-1) \geq 0$. *Applying this to each of* $x_1, \ldots, x_{10}$, *the constraints are represented by the two matrix equations* $\boldsymbol{I}_{10}\boldsymbol{x} - \mathbf{0}_{10} \geq \mathbf{0}_{10}$ *and* $-\boldsymbol{I}_{10}\boldsymbol{x} - (-\mathbf{1}_{10}) \geq \mathbf{0}_{10}$, *where* $\boldsymbol{I}_{10}$ *is the* 10×10 *identity matrix and* $\mathbf{1}_{10}$ *is the* 10×1 *vector whose every element is one. These two matrix equations may be merged to one equation of the form* $\boldsymbol{C}\boldsymbol{v} - \boldsymbol{u} \geq \mathbf{0}$ *by writing*

$$\begin{bmatrix} \boldsymbol{I}_{10} \\ -\boldsymbol{I}_{10} \end{bmatrix} \boldsymbol{v} - \begin{bmatrix} \mathbf{0}_{10} \\ -\mathbf{1}_{10} \end{bmatrix} \geq \begin{bmatrix} \mathbf{0}_{10} \\ \mathbf{0}_{10} \end{bmatrix}.$$

Unfortunately, it is easy to make a mistake when calculating a constraint or entering it to R via the matrix $\boldsymbol{C}$ *or the vector* $\boldsymbol{u}$. *Getting the sign wrong with an element of* $\boldsymbol{u}$ *(e.g., "1" instead of "−1") is one such error. Fortunately, this usually results in a warning from R that the* initial value is not in the interior of the feasible region. *If you get such a warning, and the initial guess seems to be satisfactory, it is likely that you have misspecified one or more of the constraints.*

The program that I ran appears below. It can be found in the Web site doeforglm.com as Program_4. Note that the call to constrOptim mentions the initial guess, the function to be optimised, the matrix $\boldsymbol{C}$ *and the vector* $\boldsymbol{u}$, *but it also contains two items (*`NULL` *and* `method="Nelder-Mead"`*) that have not yet been discussed. I will return to these items shortly.*

①

```
diag10 <- diag(1,10)
cmat <- rbind(diag10,-diag10)
uvec <- rep(c(0,-1),each=10)
```

Segment ① creates the identity matrix $\boldsymbol{I}_{10}$ *and the matrix* $\boldsymbol{C}$ *and vector* $\boldsymbol{u}$ *needed to define the constraints on the vector* $\boldsymbol{v}$ *of variables that are input to* constrOptim.

②

```
varbeta0hat <- function(x)
{
 a <- sum(x^2)
 v <- a/(10*a - (sum(x))^2)
 v
}
```

Segment ② defines the function varbeta0hat *(see page 33).*

③

```
start <- c(0.0026,0.0076,0.0088,0.0266,0.0450,0.0523,0.0564,0.0631,
```

```
  0.1301,0.9665)
out <- constrOptim(start,varbeta0hat,NULL,cmat,uvec,
  method="Nelder-Mead")
out
```

The starting value of $\boldsymbol{v}$ *is defined in Segment ③, then* constrOptim *is called and its output is requested.*

The output from this set of commands began with the following:

```
$par
 [1] 1.387818e-05 2.283848e-05 3.110337e-05 3.848474e-05 3.484312e-07
 [6] 5.247694e-04 3.261214e-06 2.393693e-07 3.483386e-06 9.925468e-01

$value
[1] 0.111127

$counts
function gradient
    3006       NA
```

The output gives the values of $x_1, \ldots, x_{10}$ *at which* constrOptim *has calculated that the minimum value of the variance occurs, and suggests that this minimum value is 0.111127. This minimum is clearly less than the value* $V = 0.123704$ *found earlier, so the program has been useful. At this stage, we might notice that the first nine values of* x_i *differ from 0 only by "numerical noise" (my thanks to Rolf Turner for this expression), and the last value differs from 1 by quite a small amount.*

Lots of experience with constrOptim *suggests that replacing values very close to boundaries by the actual boundary values often gives a better solution, so I tried*

```
v <- c(0,0,0,0,0,0,0,0,0,1)
varbeta0hat(v)
```

and obtained the value 0.1111111 (i.e., one-ninth, to seven decimal places). This is such an "elegant" answer that it seems likely to be correct (and substituting these values of x_i *into the function shows that the value is exactly one-ninth).*

So it is believed that the minimum value of var($\hat{\beta}_0$*) is* $\sigma^2/9$*, and occurs for the design* $x_1 = \ldots = x_9 = 0$*,* $x_{10} = 1$*. This result will be revisited in Section 2.4.3.*

Of course, the "elegance" of a solution does not justify a claim of optimality. We will see in Section 3.9 how to test whether a suspected optimum is, in fact, an optimum.

Note also in the above output from *constrOptim* that there were 3006 calls of the function *varbeta0hat* during the search for the minimum. There were zero calls of the "gradient." The gradient has not been mentioned yet, but is discussed now.

2.4.2 *Additional features of constrOptim*

It was mentioned on page 34 that one feature of *constrOptim* being used is `"method=Nelder-Mead"` . This refers to the algorithmic method used by *constrOptim* to find a minimum. The Nelder-Mead algorithm (Nelder & Mead, 1965) is an all-purpose method that does not require knowledge of the partial derivatives of $f(\boldsymbol{v})$ with respect to each of the predictor variables $v_1, \ldots, v_n$. The use of `NULL` in

```
constrOptim(start,varbeta0hat,NULL,cmat,uvec,method="Nelder-Mead")
```

tells the algorithm that the vector of "gradients" (partial derivatives)

$$\frac{\partial f}{\partial \boldsymbol{v}} = \left(\frac{\partial f}{\partial v_1}, \frac{\partial f}{\partial v_2}, \ldots, \frac{\partial f}{\partial v_n} \right)^\top$$

is not being provided by the user.

Let us examine the effect of providing the vector of partial derivatives. As there are no variables other than $x_1, \ldots, x$, then $\boldsymbol{v} = \boldsymbol{x} = (x_1, \ldots, x_n)^\top$. Given $h(\boldsymbol{v}) = h(\boldsymbol{x})$ in (2.5), it follows that

$$\frac{\partial h}{\partial x_j} = \frac{2\left(\sum_{i=1}^{10} x_i\right)\left(\sum_{i=1}^{10} x_i^2 - x_j \sum_{i=1}^{10} x_i\right)}{\left[10\sum_{i=1}^{10} x_i^2 - \left(\sum_{i=1}^{10} x_i\right)^2\right]^2} \qquad j = 1, \ldots, 10.$$

The following R function will calculate the vector of partial derivatives:

```
gradvar <- function(x)
{
 a <- sum(x^2)
 b <- sum(x)
 num <- 2*b*(a - b*x)
 denom <- (10*a - b^2)^2
 gradvector <- num/denom
 gradvector
}
```

Then I ran the following program (which, together with `gradvar`, also appears in Program_4 in the online resources).

```
start <- c(0.0026,0.0076,0.0088,0.0266,0.0450,0.0523,0.0564,0.0631,
   0.1301,0.9665)
out <- constrOptim(start,varbeta0hat,gradvar,cmat,uvec)
out
```

(with `NULL` replaced by the name of the gradient function, and `method="Nelder-Mead"` omitted). Compare this with Segment ③ on page 33.

The following output was obtained:

```
$par
 [1] 2.07716e-08 5.74861e-08 4.20601e-07 1.97813e-10 4.62986e-09
 [6] 7.22590e-08 1.06816e-08 3.91662e-09 1.73889e-07 9.75805e-01

$value
[1] 0.1111111

$counts
function gradient
     714      136
```

Note that the values of the first nine values of x are smaller numerical noise than in the previous run of *constrOptim*, but that the value of x_{10} is not as close to 1 as it was previously. However, you might guess that the minimum occurs at $x_1 = \ldots = x_9 = 0$ and $x_{10} = a$, for some value of a close to 1. Additionally the minimum value of *varbeta0hat* in the region is given as 0.1111111. Lastly, it can be seen that there were only 714 calls to the function (as compared with 3006 previously), and 136 calls to the gradient function *gradvec* before the algorithm converged. This suggests that it was worthwhile using the gradient function.

Unfortunately, there will be many occasions in this book when the function to be minimised is very complicated, and calculating its partial derivatives is not feasible. In these cases, the Nelder-Mead algorithm will be used.

2.4.3 Constrained optimisation without constrOptim

While the use of *constrOptim* generally leads to an appropriate solution, it has some disadvantages:

- Each constraint is of the form $\boldsymbol{c}^\top \boldsymbol{v} - u \geq 0$ for some vector $\boldsymbol{c}$ and scalar u. In particular, we can specify that $\delta_1 + \cdots + \delta_s \leq 1$, but we cannot require $\delta_1 + \cdots + \delta_s = 1$. I have occasionally received output from *constrOptim* where $\delta_1 + \cdots + \delta_s$ was about 0.98.
- The need to check that each constraint is satisfied at each new solution slows down *constrOptim*. This can be a disadvantage if you need to optimise a complicated function, particularly if there is a large number of variables in the problem.

A means of imposing constraints on the values of variables without formally specifying the constraints is to define the variables in such a way that they automatically satisfy those constraints. Some of the material here, together with additional discussion, can be found in Atkinson, Donev, & Tobias (2007, Section 9.5).

Select variables $z_1, z_2, \ldots$ from either a subset of $\mathbb{R}$, or from anywhere in $\mathbb{R}$ (i.e., without restriction), for input to the optimisation routine. Inside the function to be optimised, use appropriate transformations $v_1 = g_1(z_1), v_2 = g_2(z_2), \ldots$ so that the variables $v_1, v_2, \ldots$ meet the required constraints. As well as determining the appropriate transformations from $z_1, z_2, \ldots$ to $v_1, v_2, \ldots$, you need to know how to return from $v_1, v_2, \ldots$ to $z_1, z_2, \ldots$, in case you wish to modify the output from the optimisation routine to create a new guess of the optimum point.

We will consider several different situations:

1. We require $v_i \geq 0$. Then we select $z_i \in \mathbb{R}$, and calculate $v_i = z_i^2$. While the original z_i might be either $+\sqrt{v_i}$ or $-\sqrt{v_i}$, nothing will be lost if we consistently use $z_i = \sqrt{v_i}$ to find the inverse.
2. We require $0 \leq v_i \leq 1$. Again we select $z_i \in \mathbb{R}$, and then calculate either $v_i = \sin^2 z_i$ or $v_i = \cos^2 z_i$ (as, for any angle θ, $-1 \leq \sin\theta \leq 1$ and $-1 \leq \cos\theta \leq 1$). Then appropriate inverses are $z_i = \sin^{-1}(\sqrt{v_i})$ (also known as $\arcsin(\sqrt{v_i})$) and $z_i = \cos^{-1}(\sqrt{v_i})$ (also known as $\arccos(\sqrt{v_i})$), respectively. The appropriate R function names are *sin*, *cos*, *asin* and *acos*. The input to *sin* and *cos*, and the output from *asin* and *acos*, are in units of *radians*, not of *degrees*. Note that 2π radians equals $360°$. Thus `sin(pi/2)` equals $\sin 90°$; i.e., $\sin(\pi/2) = 1$.
3. We require $-1 \leq v_i \leq 1$. Here we select $z_i \in \mathbb{R}$, and then calculate either $v_i = \sin z_i$ or $v_i = \cos z_i$. The appropriate inverses are $z_i = \arcsin(v_i)$ and $z_i = \arccos(v_i)$. One could also choose $z_i \in \{z\colon 0 \leq z \leq 1\}$ and calculate $v_i = \sin(\pi z_i)$ or $v_i = \cos(\pi z_i)$, for which the inverses are $z_i = \arcsin(v_i)/\pi$ or $z_i = \arccos(v_i)/\pi$, respectively.
4. We require $a \leq v_i \leq b$ for arbitrary $b > a$. Here we select $z_i \in \mathbb{R}$, and then calculate either $v_i = [(b+a)+(b-a)\sin z_i]/2$ or $v_i = [(b+a)+(b-a)\cos z_i]/2$. The appropriate inverses are $z_i = \arcsin\{[2v_i - (b+a)]/(b-a)\}$ and $z_i = \arccos\{[2v_i - (b+a)]/(b-a)\}$.
5. We require design weights $\delta_1, \ldots, \delta_s$ to satisfy $0 < \delta_i < 1$ $(i = 1, \ldots, s)$ and $\delta_1 + \cdots + \delta_s = 1$. Here we select $z_1, \ldots, z_s$ in $\mathbb{R}$ and calculate

$$\delta_i = z_i^2 \Big/ \sum_{j=1}^{s} z_j^2 \quad (i = 1, \ldots, s). \tag{2.6}$$

The set $(z_1, \ldots, z_s)$ giving rise to a set of design weights $(\delta_1, \ldots, \delta_s)$ is not unique. For any $c \neq 0$, the set $(cz_1, \ldots, cz_s)$ will give the same $(\delta_1, \ldots, \delta_s)$. To obtain a set $(z_1, \ldots, z_s)$ that generates the set of design weights $\delta_1, \ldots, \delta_s$, note that (2.6) implies that

$$\frac{z_\ell^2}{z_m^2} = \frac{\delta_\ell}{\delta_m} \quad \text{for } \ell, m \in \{1, \ldots, s\}.$$

Let $m = s$ and arbitrarily set $z_s^2 = 1$. This gives $z_\ell^2 = \delta_\ell/\delta_s$, or

$$z_\ell = \sqrt{\delta_\ell/\delta_s} \quad (\ell = 1, \ldots, s).$$

Let *deswts* be a set of s design weights in R calculated according to (2.6). Then the command `zvec <- sqrt(deswts/deswts[s])` will calculate that set of initial weights $z_1, \ldots, z_s$ for which $z_s = 1$.

6. For case 5 above, Atkinson, Donev, & Tobias (2007, p. 131) provided an alternative transformation to produce a set of design weights. It is

$$\begin{aligned}
\delta_1 &= \sin^2 z_1 \\
\delta_2 &= \sin^2 z_2 \cos^2 z_1 \\
&\vdots \\
\delta_i &= \sin^2 z_i \prod_{j=1}^{i-1} \cos^2 z_j \quad (i = 2, \ldots, s-1) \\
&\vdots \\
\delta_s &= \prod_{j=1}^{s-1} \cos^2 z_j .
\end{aligned}$$

The design weights can be shown to add to 1 by repeated use of the result $\sin^2\theta + \cos^2\theta = 1$. Although s design weights are produced, only $(s-1)$ generators $z_1, \ldots, z_{s-1}$ are required.

To return from $(\delta_1, \ldots, \delta_s)$ to a non-unique $(z_1, \ldots, z_{s-1})$ that could have generated it, note that $\delta_1 = \sin^2 z_1$ and that $\delta_i = (\sin^2 z_i)(1 - \delta_1 - \cdots - \delta_{i-1})$ for $i = 2, \ldots, (s-1)$. From here, it is straightforward to calculate $(z_1, \ldots, z_{s-1})$.

The following two functions (both to be found in Program_5 in the online resources) will

(i) calculate $(\delta_1, \ldots, \delta_s)$ from a vector $(z_1, \ldots, z_{s-1})$, and
(ii) calculate $(z_1, \ldots, z_{s-1})$ from a vector $(\delta_1, \ldots, \delta_s)$.

The functions use the R functions *cumsum* and *cumprod* (cumulative sum and cumulative product). For $\boldsymbol{a} = (a_1, a_2, a_3, \ldots)^\top$, *cumsum(a)* gives $(a_1, a_1 + a_2, a_1 + a_2 + a_3, \ldots)^\top$, and *cumprod(a)* equals $(a_1, a_1 \times a_2, a_1 \times a_2 \times a_3, \ldots)^\top$.

(i) *Calculate* $(\delta_1, \ldots, \delta_s)$ *from* $(z_1, \ldots, z_{s-1})$

```
deswtstrig <- function(zvec)
{
 term1 <- (sin(zvec))^2
 term2 <- cumprod(1 - term1)
 deswts <- c(term1,1)*c(1,term2)
 deswts
}
```

(ii) *Calculate* $(z_1, \ldots, z_{s-1})$ *from* $(\delta_1, \ldots, \delta_s)$

```
zvectrig <- function(deswts)
{
 s <- length(deswts)
 temp <- c(0,deswts[1:(s-2)])
```

```
  ratios <- deswts[1:(s-1)]/(1-cumsum(temp))
  z <- asin(sqrt(ratios))
  z
}
```

2.4.4 Notes and examples

Here we consider the six imposed constraints from Section 2.4.3.

1. We require $v_i \geq 0$. To let a^2 be the maximum possible value for v_i, use the R command `v <- (a*runif(s))^2`. The command `runif(s)` generates s values randomly from the uniform distribution between 0 and 1. Multiplying them by a gives values between 0 and a, after which the squaring of the values gives s numbers between 0 and a^2. To return to a vector of values $\boldsymbol{z}$ from which $\boldsymbol{v}$ can be obtained by squaring each element of $\boldsymbol{z}$, use `z <- sqrt(v)`.

2. We require $0 \leq v_i \leq 1$. To generate s values of v_i randomly, I would generate values of z_i between 0 and $\pi/2$ (as $\sin^2 z_i$ can take any possible value between 0 and 1 for some z_i between 0 and $\pi/2$) using `zvec <- (pi/2)*runif(s)`. Inside the R function to be optimised, an early command should say something like `xvec <- (sin(start[i:j])^2` where the ith to jth elements of *start* contain the elements of *zvec*.

3. We require $-1 \leq v_i \leq 1$. If using $v_i = \cos z_i$, then an appropriate initial interval for z_i is $\{z : 0 \leq z \leq \pi\}$, as $\cos z_i$ takes all values between -1 and 1. The vector `zvec` can be randomly generated from `zvec <- pi*runif(s)`. If using $v_i = \sin z_i$, an appropriate interval for z_i is $\{z : -\pi/2 \leq z \leq \pi/2\}$. The vector `zvec` can be randomly generated by `zvec <- pi*(runif(s)-0.5)`. The sine transformation is a little messier than the cosine to use, so I will use only the cosine transformation. Now suppose that, once the optimisation is completed, the vector `vvec` contains the transformed values $v_1, \ldots, v_s$. Then `zvec <- acos(vvec)` will produce the pre-transformed values.

4. We require $a \leq v_i \leq b$ for arbitrary $b > a$. Produce `zvec` by `zvec <- pi*runif(s)`, and then use

   ```
   vvec <- (b+a + (b-a)*cos(zvec))/2
   ```

   to obtain the constrained values. To return from the final value of *vvec* to an appropriate *zvec*, use `zvec <- acos((2*vvec - (b+a))/(b-a))`.

5. We require design weights $\delta_1, \ldots, \delta_s$ that satisfy $0 < \delta_i < 1$ $(i = 1, \ldots, s)$ and $\delta_1 + \cdots + \delta_s = 1$. To generate an initial vector (`zvec`) of values between 0 and 1, and then produce design weights, use

   ```
   zvec <- runif(s)
   zsq <- zvec^2
   vvec <- zsq/sum(zsq)
   ```

   As stated on page 38, the command `zvec <- sqrt(vvec/vvec[s])` will produce pre-transformed values corresponding to the final constrained variables in vvec. An important advantage of this command is that it does not require the elements of *vvec* to sum to 1. Such an event might arise through the elements of *vvec* being rounded to a small number of decimal places, or if several support points are removed from, or added to, an existing design.

2.4.5 Using the function optim

Suppose that the function to be optimised has $z_1, z_2, \ldots$ as part of its input. Let the transformations that convert the z_i to constrained values form the initial commands of the function. Then the function may be optimised using the R function *optim* rather than *constrOptim*. This will generally lead to faster computations.

Invoking *optim* takes the form

```
out <- optim(start,functionname,NULL,method="Nelder-Mead")
```

where the four inputs have exactly the same meaning as they did in *constrOptim* (see Section 2.4.1). However, the matrix $\boldsymbol{C}$ and the vector $\boldsymbol{u}$ from *constrOptim* are not required here.

Example 2.4.2. *The use of transformations to impose constraints on the values of a predictor variable is now illustrated. Recall Example 2.4.1, in which the expression*

$$\frac{1}{\sigma^2} var(\hat{\beta}_0) = \sum_{i=1}^{10} x_i^2 \Bigg/ \left[\sum_{i=1}^{10} x_i^2 - (\sum_{i=1}^{10} x_i)^2\right]$$

was to be minimised, subject to $0 \leq x_i \leq 1$, $(i = 1, \ldots, 10)$.

One could generate appropriate starting values using `start <- runif(10)`. *However, there is nothing in optim to prevent alternative solutions being produced that do not lie in the appropriate domain. To prevent this, the modified function*

```
varbeta0hat <- function(z)
{
  x <- (sin(z))^2
  a <- sum(x^2)
  v <- a/(10*a - (sum(x))^2)
  v
}
```

can be used. It differs from the earlier function varbeta0hat *on page 33 through the addition of the first line* (`x <- (sin(z))^2`). *This forces each element of the vector of x-values to lie between 0 and 1, so the function optim will automatically perform constrained optimisation. The following commands in Segment ① perform the optimisation:*

①

```
set.seed(12345)
start <- (pi/2)*runif(10)
out <- optim(start,varbeta0hat,NULL,method="Nelder-Mead")
```

After optim has been run, it is necessary to transform the optimal values of z (which need not lie between 0 and 1) to the corresponding values of x that do lie between 0 and 1. The commands in Segment ② do this, and print the minimised value of the function.

②

```
xvals <- (sin(out$par))^2
xvals
out$val
```

On a single run of the program, with the seed set to 12345 so that you can verify the results if you wish to do so, the following output was obtained:

```
> xvals
 [1] 0.05388945 0.07645511 0.01682182 0.97108745 0.01397726
 [6] 0.01275970 0.08584063 0.03176969 0.81508308 0.01135588
> out$val
[1] 0.1367082
```

So the minimum value of $(1/\sigma^2)var(\hat{\beta}_0)$ *found from this single use of optim is 0.1367082.*

From Example 2.4.1, it is known that a minimum of 0.1111111 was found. Of course, this was obtained from a large number of searches for an optimum. At least two options are available. The first is to specify a value (k, say) for nsimulations *(the number of simulations to be run), and then to generate k starting values for* $(x_1, \ldots, x_{10})^\top$ *and record that design for which* $(1/\sigma^2)var(\hat{\beta}_0)$ *is a minimum. The second option is to use, as an initial guess, the solution 0.0026, 0.0076, 0.0088, 0.0266, 0.0450, 0.0523, 0.0564, 0.0631, 0.1301, and 0.9665 that was found from 20 million simulations without any attempt at optimisation (see page 31).*

Option 1 *The following program was used:*

```
set.seed(12345)
nsimulations <- 1000
min <- 10
for (i in 1:nsimulations)
{
 start <- (pi/2)*runif(10)
 out <- optim(start,varbeta0hat,NULL,method="Nelder-Mead")
 if(out$val < min) {min <- out$val
    solution <- out$par}
}
 xvals <- (sin(solution))^2
 xvals
 min
```

This gave the output

```
> xvals
 [1] 1.875476e-05 3.420228e-07 2.451684e-05 5.418596e-05 3.006119e-05
 [6] 1.170911e-06 1.644166e-06 8.037705e-01 3.853950e-06 2.021875e-08
> min
[1] 0.1111152
```

Option 2 *This program was used:*

```
start <- sqrt(asin(c(0.0026,0.0076,0.0088,0.0266,0.0450,0.0523,
  0.0564,0.0631,0.1301,0.9665)))
out <- optim(start,varbeta0hat,NULL,method="Nelder-Mead")
xvals <- (sin(out$par))^2
xvals
out$val
```

Note that the tentative solution consisted of x-values, whereas the input to the function varbeta0hat *needs to consist of z-values. So the first of the commands in Option 2 performs a conversion from x-values to z-values. The result of this single run of* optim *was*

```
> xvals
 [1] 8.158784e-05 2.111352e-04 3.944535e-05 4.932786e-06 1.701881e-04
 [6] 6.968760e-08 2.907663e-06 5.089606e-05 4.432624e-05 9.381707e-01
> out$val
[1] 0.1111271
```

It is fairly clear from the results of either Option 1 or Option 2 that, as found in Subsections 2.4.1 and 2.4.2, the optimal solution has $x_1 = x_2 = \ldots = x_9 = 0$ *and* $x_{10} = a$*, for some still-to-be-determined value of a satisfying* $0 < a \leq 1$*. (The value of a must not equal 0, or else all 10 values of* x_i *would be equal. One cannot estimate the slope or y-intercept of a line with just one x-value.)*

In fact, if one substitutes $x_1 = \ldots = x_9 = 0$ *and* $x_{10} = a\,(> 0)$ *into* (2.5)*, one obtains an answer of* $1/9 = 0.1111111$ *irrespective of the value of a. It seems that there is not a unique set of values of* $x_1, \ldots, x_{10}$ *that minimise var*$(\hat{\beta}_0)$.

In such circumstances, one could recognise the existence of $\hat{\beta}_1$ *in the estimation of the parameters. Simply noting that var*$(\hat{\beta}_1)$ *is minimised when the x-values are at the extremities would suggest that* $a = 1$ *should be chosen. Another approach would be to ask what is the best design for minimising var*$(\hat{\beta}_0)$ *when* β_1 *is in the model. This will be investigated in Sub-section 3.7.5.*

2.4.6 Initial values for optimisations

It will be apparent from the examples already seen that an optimisation routine is not guaranteed to find an optimal value of a function. It depends very much on the initial value entered into the routine. This should not be surprising. If you were to be dropped at random at some point in the Himalayan mountains and you were able to reach the peak of the highest mountain in sight, would you be at the top of Mount Everest? Not necessarily! It would depend very much on whether you were dropped close to Mount Everest.

If you have some knowledge of roughly what is the optimal design (the values of the support points and of their design weights), then certainly you should use that knowledge to suggest a starting value for the optimisation routine. That knowledge may arise because you have already performed an optimisation routine for a similar value of the parameter vector $\boldsymbol{\beta}$. However, if you have no useful knowledge of a starting value, my preference is to generate some permissible starting values of $\boldsymbol{x}_1, \ldots, \boldsymbol{x}_s$ and $(\delta_0, \ldots, \delta_s)^\top$ at random, find the 'optimal' design from amongst those found so far, then use minor adjustments to this design as a set of starting values for further searches, and so on ... This will be demonstrated numerous times in subsequent chapters.

2.5 Numerical integration

In the study of calculus, the topic of differentiation is generally followed by integration (occasionally called "anti-differentiation"). This is sometimes the first topic where scientists do not study the mathematics learnt by mathematicians and statisticians, so a brief consideration will be given to it and to an extension of it. The expression

$$\int_a^b f(x)\,dx$$

is read as "the integral from $x = a$ to $x = b$ of the function $f(x)$ with respect to x," and it can be interpreted as the area in the (x, y)-plane that is bounded by the curve $y = f(x)$, the straight line $y = 0$, and the two straight lines $x = a$ and $x = b$. This is the area of the region that is shaded in Figure 2.3.

For many functions $f(x)$, there is a mathematical expression for the value of the integral in (2.5), and this value can either be calculated exactly or to as much precision as is required. However, there are other functions for which such a value cannot be obtained, either because of the nature of the function or because there is no simple way to write the function in terms of x. For these latter functions, the integral must be evaluated by approximate methods, generally known as *numerical integration.* A well-known example of the results of numerical integration is the standard normal tables that appear at the back of most statistics textbooks; all the probabilities were originally calculated by numerical integration.

Basic methods of numerical integration, such as *Simpson's rule,* are often taught in high school mathematics. Simpson's rule says that

$$\int_a^b f(x)\,dx \approx \frac{b-a}{6}\left[f(a) + 4f(\frac{a+b}{2}) + f(b)\right], \tag{2.7}$$

which allows the definite integral of a function to be approximated by use of the values of the function at several values of x.

If $f(x) = c_0 + c_1 x + c_2 x^2 + c_3 x^3$ (with at least one c_i being nonzero), the expression on the right of (2.7) is in fact the exact value of the integral. For more complicated functions, the expression is just an approximation to the

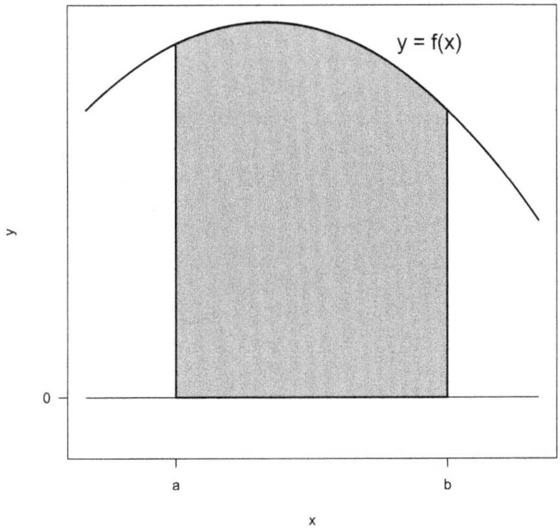

Figure 2.3 *The area of the shaded region is the value of the integral* $\int_a^b f(x)\,dx$.

value of the integral. The approximation can be improved by splitting the interval from $x = a$ to $x = b$ into several mutually exclusive and exhaustive sub-intervals, applying the result to each sub-interval, and then adding the individual approximations together to give an overall answer.

Suppose that $[a, b]$ is divided into n sub-intervals of equal width. Let x_{2i-2}, x_{2i-1} and x_{2i} represent the least value, midpoint and greatest value of x in the ith interval ($i = 1, \ldots, n$), and let $h = (b - a)/(2n)$ represent the distance between two consecutive values of x. Note that, for all but the last interval, the greatest value of an interval is also the least value of the next interval. It then follows that (2.7) can be generalised to

$$\int_a^b f(x)\,dx \approx \frac{h}{3}[f(x_0) + 4f(x_1) + 2f(x_2) + 4f(x_3) + 2f(x_4) + \cdots + 2f(x_{2n-2}) + 4f(x_{2n-1}) + f(x_{2n})]. \tag{2.8}$$

Example 2.5.1. *Consider the function*

$$f(x) = e^{-x}[2 + \sin(ux) + \sin(vx)]/2,$$

where u and v are constants. The integral

$$I(u, v) = \int_0^1 f(x)\,dx \tag{2.9}$$

can be shown to equal

$$I(u,v) = 1 - e^{-1} + \left[\frac{u - e^{-1}(u\cos u + \sin u)}{1+u^2} + \frac{v - e^{-1}(v\cos v + \sin v)}{1+v^2}\right]/2,$$

so strictly speaking there is no need to evaluate the integral numerically. However, it has been chosen as an example of an integral that is not trivial.

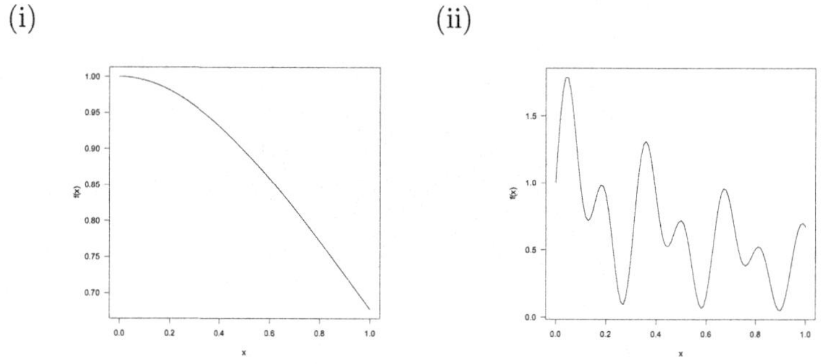

Figure 2.4 *Plots of* $f(x)$ *in* (2.5.1) *vs.* $x \in (0,1)$ *for (i)* $u = 1$, $v = 1$ *and (ii)* $u = 40$, $v = 20$.

Figure 2.4 contains plots of $f(x)$ *vs.* x *for (i)* $u = 1$, $v = 1$ *and (ii)* $u = 40$, $v = 20$. *The plot in (i) is smooth, while that in (ii) is quite "wiggly." The smoothness of the curve in (i) allows Simpson's rule in* (2.8) *to work very well even when the interval from 0 to 1 is divided into just two sub-intervals, each of width 0.5. Simpson's rule gives a value of 0.8777845, which agrees well with the exact answer of 0.8779576 (to seven decimal places). In case (ii), the wiggliness means that Simpson's rule requires a large number of sub-intervals in order for the curve in any sub-interval to be smooth, and more sub-intervals do not necessarily give an answer closer to the true value. Table 2.1 contains the answers given by Simpson's rule for varying number of sub-intervals of equal width. The approximation for four sub-intervals is not closer to the true value of 0.6683673 than the value for two sub-intervals, but a good approximation is achieved if the number of sub-intervals is increased sufficiently.*

Number of sub-intervals of equal width					
2	4	10	20	40	50
0.7577521	0.4545241	0.6202149	0.6710641	0.6684748	0.6684094

Table 2.1 *Approximations given by the application of Simpson's rule using varying numbers of equal-width segments to the integral in* (2.9) *for* $u = 40$ *and* $v = 20$. *The true value is 0.6683673.*

The formula in (2.8) can be written more generally as

$$\int_a^b f(x)\,dx \approx \sum_{i=0}^{2n} w_i f(x_i),$$

where w_i is the "weight" associated with the ith x-value (or "abscissa"). The use of this formula requires $f(x)$ to be evaluated $(2n+1)$ times.

Simpson's rule can be extended to integrals of more than one variable. Consider the double integral

$$\int_a^b \int_c^d f(x,y)\,dy\,dx,$$

which is the volume of a solid in three dimensions. An understanding of what this solid represents is not necessary to proceed further in this book. Suppose that the intervals from a to b and from c to d are divided into n and m subintervals of lengths $(b-a)/n$ and $(d-c)/m$, respectively. Then the Simpson's rule approximation to the above integral is of the form

$$\sum_{i=0}^{2n}\sum_{j=0}^{2m} w_{ij} f(x_i, y_j),$$

where w_{ij} represents a weight. The function $f(x,y)$ must be evaluated a total of $(2n+1)(2m+1)$ times.

Some of the calculations in later chapters will require use of numerical integration. Because the value of the function $f(\cdot)$ will be calculated many times, it is necessary to try to make the R program as efficient as possible in order for the computations to be done in an acceptable time period.

2.6 Conclusion

This chapter has given mathematical results needed in the search for an optimal design for a GLM. This book uses the freely available software R to perform the necessary computations. Examples have been given of some of the programs that will be used.

The function *constrOptim* may be used to achieve constrained optimisation. However, it is recommended that the function *optim* be used in conjunction with transformations whose results give input variables that satisfy the required constraints.

The concept of numerical integration has been introduced. It will be used in Sections 4.8 and 5.4 and in Chapter 7.

Chapter 3

The Theory Underlying Design

3.1 Introduction

Chapter 3 provides most of the background material for the rest of the book. It describes various criteria that are in use to select optimal designs, and emphasises the general equivalence theorem, which enables a researcher to check whether a design is actually optimal. A problem is discussed that arises if the total number of observations is small. The material in the chapter is illustrated by numerous examples.

3.2 Notation

The notation to be used in the remaining chapters is now introduced.

Recall that both general and generalized linear models use a linear predictor, η, to model the mean of an observation on a random variable, Y. Write m for the number of mathematically independent explanatory variables used in the linear combination of parameters, η. Denote these variables by $x_1, \ldots, x_m$, and write $\boldsymbol{x} = (x_1, \ldots, x_m)^\top$. Let there be p parameters in η. These will generally be denoted by $\beta_0, \beta_1, \ldots, \beta_{p-1}$, and will form the elements of the $p \times 1$ vector $\boldsymbol{\beta} = (\beta_0, \ldots, \beta_{p-1})^\top$. In η, the coefficient of each β_i will be a function (called a "regressor") of one or more of $x_1, \ldots, x_m$. Denote by $f_i(\boldsymbol{x})$ the function that is the coefficient of β_i $(i = 0, \ldots, p-1)$, and write

$$\boldsymbol{f}(\boldsymbol{x}) = (f_0(\boldsymbol{x}),\ f_1(\boldsymbol{x}), \ldots, f_{p-1}(\boldsymbol{x}))^\top .$$

It follows that

$$\eta = f_0(\boldsymbol{x})\beta_0 + \cdots + f_{p-1}(\boldsymbol{x})\beta_{p-1} = \boldsymbol{f}^\top(\boldsymbol{x})\boldsymbol{\beta}.$$

3.3 Designing an experiment

3.3.1 *Exact and approximate designs*

Designing an experiment requires (i) the selection of values of $\boldsymbol{x}$ at which observations on Y will be made, and (ii) a determination of how many independent observations on Y will be made at each $\boldsymbol{x}$.

If $\boldsymbol{x}$ is one of the points used in the design, it will be called a *support point* of the design. Denote by s the number of support points. Let n_i (> 0) represent the number of independent observations made at the ith support point, $\boldsymbol{x}_i$.

Write $N = n_1 + \cdots + n_s$. Then $\delta_i = n_i/N$ is the proportion of observations taken at $\boldsymbol{x}_i$. Clearly $\delta_i > 0$ $(i = 1, \ldots, s)$ and $\delta_1 + \cdots + \delta_s = 1$. The quantities $\delta_1, \ldots, \delta_s$ are called the *design weights* of the s support points.

It is convenient to describe a design ξ by the collection of support points $\boldsymbol{x}_1, \ldots, \boldsymbol{x}_s \in \mathbb{R}^m$ and their corresponding design weights. We write

$$\xi = \left\{ \begin{array}{cccc} \boldsymbol{x}_1 & \boldsymbol{x}_2 & \ldots & \boldsymbol{x}_s \\ \delta_1 & \delta_2 & \ldots & \delta_s \end{array} \right\}. \tag{3.1}$$

Example 3.3.1. *Recall from page 18 that Design 2 was intended for* $m = 1$ *explanatory variable* (x_1, *written as* x *for simplicity). It had* $s = 2$ *support points, at* $x = 0$ *and* $x = 1$, *and the 10 observations were allocated five each to the support points. Then we could write*

$$\xi_2 = \left\{ \begin{array}{cc} 0 & 1 \\ 5/10 & 5/10 \end{array} \right\}$$

to represent Design 2. This design was intended specifically for $N = 10$ *observations. If we had designed the experiment for* $N = 12$ *observations, the design would have had six observations at each support point, and the design weights would have been written as* 6/12 *and* 6/12.

Designs that are created for a specific value of N are called *exact designs*, and the design weights are left as fractions (with no cancellation performed) to indicate the value of N. That is, the design weights of ξ_2 are not written as 0.5 for each of $x = 0$ and $x = 1$ when the design is an exact design.

In an *approximate design*, also called a *continuous design*, the design weights are usually given as decimals (but not always: design weights that are recurring decimals are best left as fractions; e.g., 1/3). Approximate designs represent idealised designs that are rarely exactly achievable. For example, a design with three support points and design weights 0.25, 0.5 and 0.25 can only be achieved if N is a multiple of four. If the weights are (say) 0.233, 0.534 and 0.233, an exact design is not achievable for small values of N.

3.3.2 Constructing an exact design from an approximate design

Approximate designs represent a "target" at which we can aim when producing an exact design. It is assumed that the values of the support points $\boldsymbol{x}_i$ $(i = 1, \ldots, s)$ are left unchanged in the exact design. When the value of N has been established, then we can investigate designs for which $n_i \approx N\delta_i$ $(i = 1, \ldots, s)$. Unfortunately, the n_i produced in this way are rarely integers. If nonintegral values are rounded to the nearest integer, it often happens that the resulting sample sizes do not sum to N. Various methods have been suggested to overcome this difficulty. Pukelsheim (1993, Chapter 12) describes an efficient apportionment that allocates sample sizes n_i $(i = 1, \ldots, s)$ to the various support points. These sample sizes satisfy $\sum_i n_i = N$, and the method of apportionment is shown by Pukelsheim to have optimal properties. Additionally, the method works if the recording of the values of the δ_i to a specified number of decimal places leads to $\sum \delta_i$ not equalling 1 exactly.

In some situations, the efficient apportionment does not give a unique set of sample sizes $(n_1, \ldots, n_s)$. This occurs when several quantities being ranked from smallest to largest have tied values, and the various apportionments correspond to different tied values being selected by the procedure. Pukelsheim (1993, Exhibit 12.2, p. 310) considers the situation where $s = 3$ and $\delta_1 = 1/6$, $\delta_2 = 1/3$ and $\delta_3 = 1/2$. Pukelsheim shows that three apportionment sets occur when $N = 6k + 1$ or $N = 6k + 2$ for $k = 1, 2, 3, \ldots$, but that a unique apportionment occurs for other values of N. For example, when $N = 8$, the apportionments $(2, 3, 3)$, $(2, 2, 4)$ and $(1, 3, 4)$ may all be used but, when $N = 9$, only $(2, 3, 4)$ is indicated.

Program_6 in doeforglm.com will perform an efficient apportionment when given the values of N and $\delta_1, \ldots, \delta_s$. Where more than one efficient apportionment exists, the program will tell you that more than one exists and select one at random. To see more than one such apportionment, run the program several times.

Example 3.3.2. *Suppose that $s = 5$ and $\delta_1 = 0.246$, $\delta_2 - 0.301$, $\delta_3 = 0.109$, $\delta_4 = 0.125$ and $\delta_5 = 0.219$. For $N = 18$, Program_6 gives the unique efficient apportionment*

```
4 5 3 2 4
```

as compared to the result

```
4 5 2 2 4
```

(which does not sum to 18) that occurs with simple rounding of each $N\delta_i$.

An exact design constructed from an approximate design will not be optimal because the proportions of observations taken at the support points have been altered slightly from the optimal values. However, it is likely to be "close to optimal." A comparison of an approximate design with its exact counterpart will be considered in an example in Section 4.7.

3.3.3 Constructing an exact design directly

One can also construct a "close to optimal" exact design by modifying the function that is to be optimised by either *optim* or *constrOptim*. One first selects the number of support points desired, remembering that it is necessary to have $s \geq p$. These will have weights $1/s$ each, although it is possible that the optimisation procedure may select a support point more than once, thereby giving it a greater weight. This occurrence will be illustrated in Section 4.7.

The modification of the function is straightforward. Instead of giving $\boldsymbol{v} = (x_{11}, \ldots, x_{1s}, x_{21}, \ldots, x_{2s}, \ldots, x_{ms}, \delta_1, \ldots, \delta_m)$ as the argument of the function, one gives $\boldsymbol{v} = (x_{11}, \ldots, x_{1s}, x_{21}, \ldots, x_{2s}, \ldots, x_{ms})$ as the argument, and specifies the values of $\delta_1, \ldots, \delta_m$ either in the global environment or inside the program where they cannot be altered by the optimisation routine.

Example 3.3.3. *A function called detinfomat appears twice in the program beginning on page 106. The function has $\boldsymbol{z} = (z_1, \ldots, z_{(m+1)s})$ as its argument. Inside the unmodified function, $z_1, \ldots, z_{ms}$ are transformed to values*

$x_{11}, \ldots, x_{1s}, x_{21}, \ldots, x_{2s}, \ldots, x_{ms}$ *that satisfy the criteria* $-1 \leq x_{ij} \leq 1$ $(i = 1, \ldots, m;\ j = 1, \ldots, s)$*, while* $z_{ms+1}, \ldots, z_{(m+1)s}$ *are transformed to* $\delta_1, \ldots, \delta_s$*, with* $0 \leq \delta_i \leq 1$; $\sum_i \delta_i = 1$. *From these constrained values, the information matrix* $\boldsymbol{M}(\xi, \boldsymbol{\beta})$ *is calculated, followed by* $-det[\boldsymbol{M}(\xi, \boldsymbol{\beta})]$. *This is achieved by the following part of the function:*

```
lim1 <- m*s
lim2 <- (m+1)*s
detinfomat <- function(variables)
{
 xmat <- matrix(cos(pi*variables[1:lim1]),m,s,byrow=T)
 zvec <- variables[(lim1+1):lim2]
 deswts <- zvec^2
 deltavec <- deswts/sum(deswts)
 :
```

If, instead, the design weights are to be $1/s$ *for each of* s *support points, the argument of the function,* variables*, should have only* ms *elements. The start of the program may be reduced to*

```
deltavec <- rep(1/s,s)
detinfomat <- function(variables)
{
 xmat <- matrix(cos(pi*variables),m,s,byrow=T)
 :
```

A numerical example appears in Sub-section 4.7.

An alternative method of constructing a "close to optimal" exact design involves the use of a program to construct optimal Bayesian designs, and is deferred to Section 7.4.

3.4 Selecting the support points

3.4.1 Thinking about criteria for selection

Most work on general and generalized linear models considers the analysis of data that are thought to follow a particular model that involves m explanatory variables $x_1, x_2, \ldots, x_m$ and possibly some functions of them (e.g., $x_1 x_2$). We postulate a linear model that says that some function of a distribution parameter is equal to a linear combination of the unknown model parameters $\beta_0, \ldots, \beta_{p-1}$. The aim of the analysis is to estimate the values of $\beta_0, \ldots, \beta_{p-1}$, and to test hypotheses about their true values, with the aim of refining the linear combination.

In most textbooks, the data to be analysed are simply presented to us. The researcher believed that the response variable might be affected by the values of the explanatory variables. How were the values of the explanatory variables selected? *Does it matter how they were selected?* Yes, it does matter.

A situation already considered is where the mean of the response variable Y depends on the value of an explanatory variable, x, and the following relationship is specified:

- $g(\mu) = \mu = \eta = \beta_0 + \beta_1 x$;
- the variance of Y takes the same value, σ^2, for all values of x;
- for a given value of x, the distribution of Y is normal.

If necessary, x has been scaled so that it takes values between 0 and 1, and we assume that we can afford to take observations on Y at $n = 10$ values of x. Where should those values be?

To answer this question requires the formulation of some criteria about what interests us in the model. We shall shortly see various commonly used criteria but, for the moment, we shall assume that we simply want to estimate the value of the slope (β_1) "as well as possible."

What does this mean? Mathematical theory tells us that the LS and ML estimator of β_1 is *unbiased*. This means that if many samples are taken and the estimate of β_1 is calculated for each sample, the average value will theoretically equal the true value of β_1. This will occur irrespective of what set of ten points we use. However, the *spread* of these estimates around the true β_1 does depend on the particular set of ten points chosen, and we would prefer to use the design that gives the smallest spread (the smallest standard error of the estimator).

Recall Design 1 and Design 2 from Example 1.5.1, where we considered just the variance of $\hat{\beta}_1$ for the standard straight line regression. Program_7 in the online resources generates samples of 10 observations from the two designs. It uses the arbitrary values $\beta_0 = 2$, $\beta_1 = 0.7$ and $\sigma = 0.4$ for purposes of demonstration. By simple changes to the program, you can alter the number of simulations, and the true values of β_0, β_1 and σ. To eliminate any concern that differences between the results from the two designs might depend on the random errors associated with the observations, the same errors are used for both designs in any single run of the program. A seed has been specified in the program so that, if you wish, you can emulate the program exactly and check that you get the same results as are presented here. If you are not interested in checking this, just omit the command line that sets the seed.

The following results were obtained from 1000 simulations of conducting the experiment using each design. Design 1 (the "equispaced" design) has 10 independent observations, one each at $x = 0, 1/9, 2/9, \ldots, 8/9, 9/9$, while Design 2 (the "extremities" design) has five independent observations at $x = 0$ and five independent observations at $x = 1$. Table 3.1 contains the average and standard deviation of the 1000 estimates of β_1 obtained from each design, and Figure 3.1 shows boxplots of the two sets of 1000 estimates of β_1.

The value of β_1 used in the simulations was 0.7. The average values of $\hat{\beta}_1$ from the two designs are both very close to 0.7. The standard deviation of the estimates from the "extremities" design is about 65% of the corresponding value from the "equispaced" design.

	Design	
	Equispaced	Extremities
Mean	0.7137061	0.7004807
Std. dev.	0.3948611	0.2554848

Table 3.1 *Means and standard deviations of 1000 estimates of* $\beta_1 = 0.7$ *from two different designs.*

The results of these simulations are consistent with the theoretical results obtained for the variance of $\hat{\beta}_1$ under Designs 1 and 2 in Example 1.5.1, where $\text{var}(\hat{\beta}_1)$ was $54\sigma^2/55$ for Design 1 and $2\sigma^2/5$ for Design 2. With $\sigma = 0.4$, this suggests standard errors of $\sqrt{(54 \times 0.4^2/55)} = 0.3963$ for Design 1 and $\sqrt{(2 \times 0.4^2/5)} = 0.2530$ for Design 2.

The difference in the distributions of the estimates from the two designs is shown clearly in Figure 3.1. Using the "extremities" design makes it more likely that the estimate of β_1 will be close to the true value. This suggests that the "extremities" design is to be preferred to the "equispaced" design.

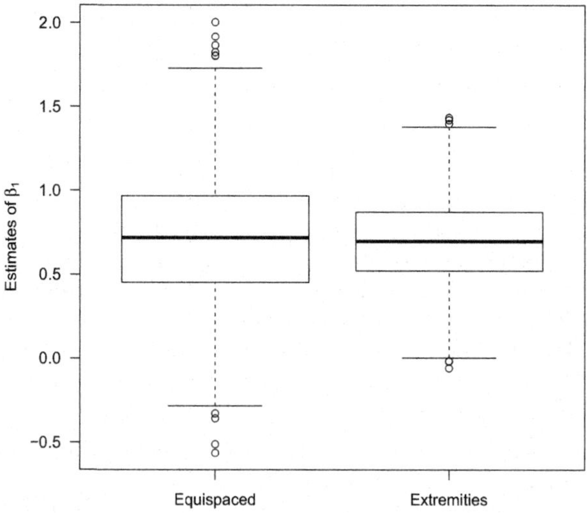

Figure 3.1 *Boxplots of 1000 estimates of* β_1*, from the designs with support points equispaced over, and at the extremities of, the domain of* x*.*

Q: Could this result be influenced by the values chosen for β_0*,* β_1 *or* σ*?*

A: No. The results agree very well with theory. But you don't have to trust this statement. Change the values of β_0*,* β_1 *and/or* σ *in Program_7, then run the* *program and check what happens. You can change the number of simulations, too, if you wish — but don't make it too small, or sampling error might distort the results.*

There is one situation where the "extremities" design might not be appropriate. This is when the relationship between the response variable and the predictor variable, x, is *not* a straight line. If the true relationship is quadratic, or cubic, or ... , this cannot be detected by a design with only two distinct values of x. Figure 3.2 shows the unique straight line passing through two points, and also shows two of the infinitely many parabolas (curves that have a quadratic relationship between y and x) that also pass through these two points. The "extremities" design cannot tell you which of these three relationships (or others) is appropriate.

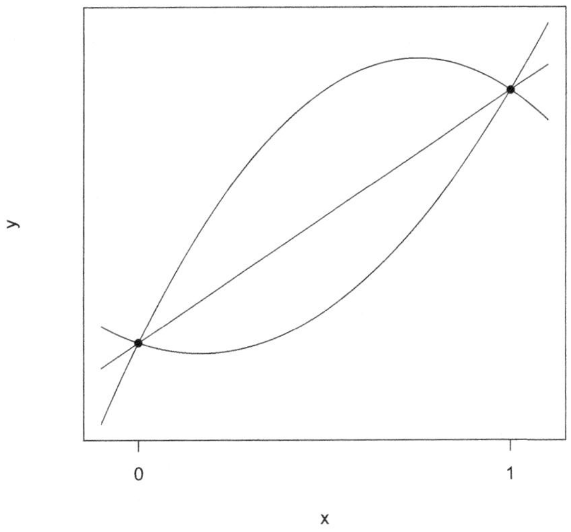

Figure 3.2 *The unique straight line and two of the infinitely many parabolas that pass through the indicated two points.*

If you are sure that the true relationship between μ and x is a straight line, use the "extremities" design. If there is some doubt about whether it is a straight line, use a design that can detect possible curvature. You might do this by reducing the number of observations at $x = 0$ and $x = 1$ to four each, and taking two observations at $x = 0.5$. *Is this the "best" design?* We will consider this question in Example 3.7.3.

Choosing between various designs by considering the precision of the estimators that we obtain is effectively the same as looking at the *widths* of the confidence intervals for β_0 and β_1. Under the usual statistical assumptions underlying the general linear model, the vector of estimators, $\hat{\boldsymbol{\beta}} = (\hat{\beta}_0, \hat{\beta}_1)^\top$, has a bivariate normal distribution with mean $\boldsymbol{\beta} = (\beta_0, \beta_1)^\top$ and covariance matrix $\sigma^2 \boldsymbol{V}$, where $\boldsymbol{V} = [\boldsymbol{F}^\top \text{diag}(n_1, \ldots, n_s)\boldsymbol{F}]^{-1}$ (see page 16), written as

$$\hat{\boldsymbol{\beta}} \sim \text{N}_2(\boldsymbol{\beta}, \sigma^2 \boldsymbol{V}). \tag{3.2}$$

Let v_{ij} represent the (i, j) element of $\boldsymbol{V}$, $i = 1, 2$; $j = 1, 2$. It follows that $\beta_0 \sim N(\beta_0, \sigma^2 v_{11})$ and $\beta_1 \sim N(\beta_0, \sigma^2 v_{22})$, which implies that

$$\frac{\hat{\beta}_0 - \beta_0}{\sqrt{(\sigma^2 v_{11})}} \sim \text{N}(0, 1) \quad \text{and} \quad \frac{\hat{\beta}_1 - \beta_1}{\sqrt{(\sigma^2 v_{22})}} \sim \text{N}(0, 1).$$

As the value of σ^2 is unknown, we would replace it by the Residual MS (ResMS) from the ANOVA table of the analysis. The statistic ResMS has $(N - 2)$ degrees of freedom. This leads to the results

$$\frac{\hat{\beta}_0 - \beta_0}{\sqrt{(\text{ResMS}\, v_{11})}} \sim t_{N-2} \quad \text{and} \quad \frac{\hat{\beta}_1 - \beta_1}{\sqrt{(\text{ResMS}\, v_{22})}} \sim t_{N-2}.$$

Then $100(1 - \alpha)\%$ confidence intervals for β_0 and β_1 are respectively

$$\left(\hat{\beta}_0 - t_{N-2,1-\alpha/2}\sqrt{(\text{ResMS}\, v_{11})},\ \hat{\beta}_0 + t_{N-2,1-\alpha/2}\sqrt{(\text{ResMS}\, v_{11})}\right)$$

and

$$\left(\hat{\beta}_1 - t_{N-2,1-\alpha/2}\sqrt{(\text{ResMS}\, v_{22})},\ \hat{\beta}_1 + t_{N-2,1-\alpha/2}\sqrt{(\text{ResMS}\, v_{22})}\right),$$

where $t_{N-2,1-\alpha/2}$ represents the $100(1 - \alpha/2)\%$ quantile of the t_{N-2} distribution; e.g., $t_{12,0.95} = 1.782$.

We have already determined the designs for $N = 10$ which individually minimise the widths of these two intervals.

If these intervals are both plotted on the (β_0, β_1) plane, they mark out a rectangular region. See Example 3.4.1 and Figure 3.3. However, the formation of this region takes no account of the fact that the estimators $\hat{\beta}_0$ and $\hat{\beta}_1$ are correlated. Knowledge of the correlation should help to identify values of $(\hat{\beta}_0, \hat{\beta}_1)$ that are comparatively unlikely to occur. Consequently, we should be able to replace the rectangular region by an alternative region that is more informative. This region is called a *confidence region.* We will now derive a confidence region for (β_0, β_1). For this case, where $p = 2$, the region will be an ellipse. A confidence region for $p = 3$ model parameters will be an ellipsoid (of the general shape of a Rugby football or an Australian or American football). A confidence region for $p > 3$ parameters is called a *hyperellipsoid;* this name is also used as an all-purpose name for the region for any value of p.

From (3.2), it follows that

$$(\hat{\boldsymbol{\beta}} - \boldsymbol{\beta})^\top (\sigma^2 \boldsymbol{V})^{-1} (\hat{\boldsymbol{\beta}} - \boldsymbol{\beta}) \sim \chi^2_2.$$

Dividing the expression on the LHS by 2, and replacing the unknown value of σ^2 by ResMS gives the result

$$(\hat{\boldsymbol{\beta}} - \boldsymbol{\beta})^\top (\text{ResMS}\, \boldsymbol{V})^{-1}(\hat{\boldsymbol{\beta}} - \boldsymbol{\beta})/2 \sim F_{2,N-2}.$$

As $\boldsymbol{V}^{-1} = \boldsymbol{F}^\top \text{diag}(n_1, \ldots, n_s)\boldsymbol{F}$, it follows that $W \sim F_{2,N-2}$, where

$$W = (\hat{\boldsymbol{\beta}} - \boldsymbol{\beta})^\top [\boldsymbol{F}^\top \text{diag}(n_1, \ldots, n_s)\boldsymbol{F}](\hat{\boldsymbol{\beta}} - \boldsymbol{\beta})/(2 \times \text{ResMS}).$$

Hence a $100(1-\alpha)\%$ joint confidence region for $(\beta_0,\, \beta_1)$ is given by $W \leq F_{2,N-2;\,1-\alpha}$, where $F_{2,N-2;\,1-\alpha}$ is the $100(1-\alpha)\%$ quantile of the $F_{2,N-2}$ distribution.

Example 3.4.1. *Suppose that the "equispaced" design for $N = 10$ is used, and an experiment is run in which the value of Y is observed at each of the 10 values of x. Then, from page 18,*

$$\boldsymbol{V} = [\boldsymbol{F}^\top \text{diag}(n_1, \ldots, n_s)\boldsymbol{F}]^{-1} = \frac{1}{81}\begin{bmatrix} 19 & -27 \\ -27 & 54 \end{bmatrix},$$

and $v_{11} = 19/55$ and $v_{22} = 54/55$.

Further suppose that the statistical analysis produced the results $\hat{\beta}_0 = 0.74$, $\hat{\beta}_1 = 1.28$ and ResMS $= 4.35$. If individual 95% equal-tailed confidence intervals are constructed for β_0 and β_1, then $t_{10-2,1-0.025} = t_{8,0.975} = 2.306$ is used, and the confidence intervals are

$$\begin{aligned}
\beta_0: \quad & \left(\hat{\beta}_0 - t_{8,0.975}\sqrt{(\text{ResMS}\ v_{11})},\ \hat{\beta}_0 + t_{8,0.975}\sqrt{(\text{ResMS}\ v_{11})}\right) \\
= \quad & (0.74 - 2.306\sqrt{(4.35 \times 19/55)},\ 0.74 + 2.306\sqrt{(4.35 \times 19/55)}) \\
= \quad & (-2.087,\ 3.567), \\
\beta_1: \quad & \left(\hat{\beta}_1 - t_{8,0.975}\sqrt{(\text{ResMS}\ v_{22})},\ \hat{\beta}_1 + t_{8,0.975}\sqrt{(\text{ResMS}\ v_{22})}\right) \\
= \quad & (1.28 - 2.306\sqrt{(4.35 \times 54/55)},\ 1.28 + 2.306\sqrt{(4.35 \times 54/55)}) \\
= \quad & (-3.486,\ 6.046).
\end{aligned}$$

As $F_{2,8;0.95} = 4.4590$, the joint confidence region for $(\beta_0,\, \beta_1)$ is

$$(\hat{\boldsymbol{\beta}} - \boldsymbol{\beta})^\top [\boldsymbol{F}^\top \text{diag}(n_1, \ldots, n_s)\boldsymbol{F}](\hat{\boldsymbol{\beta}} - \boldsymbol{\beta})/(2 \times 4.35) \leq 4.4590$$

or, equivalently,

$$(\boldsymbol{\beta} - \hat{\boldsymbol{\beta}})^\top [\boldsymbol{F}^\top \text{diag}(n_1, \ldots, n_s)\boldsymbol{F}](\boldsymbol{\beta} - \hat{\boldsymbol{\beta}}) \leq 2 \times 4.35 \times 4.4590.$$

This consists of the boundary and interior of the ellipse given by

$$(\boldsymbol{\beta} - \hat{\boldsymbol{\beta}})^\top [\boldsymbol{F}^\top \text{diag}(n_1, \ldots, n_s)\boldsymbol{F}](\boldsymbol{\beta} - \hat{\boldsymbol{\beta}}) = 38.79304.$$

The centre of the ellipse is at $(\hat{\beta}_0,\, \hat{\beta}_1)$.

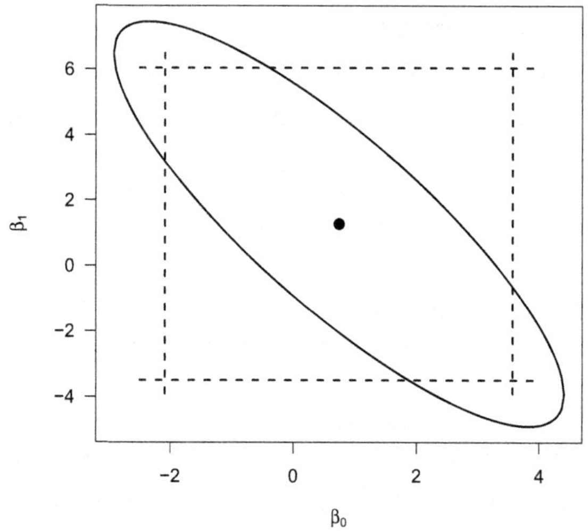

Figure 3.3 *The boundaries (dotted lines) of the individual 95% confidence intervals for β_0 and β_1, and the joint 95% confidence region for (β_0, β_1) from Example 3.4.1. The centre of the ellipse is marked by a solid point.*

Figure 3.3 shows the rectangular region whose boundaries are created by the individual confidence intervals for β_0 and β_1. The figure also shows the ellipse that forms the boundary of the joint confidence region.

It can be seen that there are many values of (β_0, β_1) that lie within both individual confidence intervals but do not lie in the joint confidence region. There are also some values of (β_0, β_1) that lie in the joint confidence region but are not within both individual confidence intervals.

The joint confidence region makes use of the fact that $\hat{\beta}_0$ and $\hat{\beta}_1$ are negatively correlated (correlation $= -0.843$). This means that it is very unlikely that we would get a "much higher than average" estimate of β_0 in conjunction with a "much higher than average" estimate of β_1, or a "much lower than average" estimate of β_0 in conjunction with a "much lower than average" estimate of β_1. This lets us regard some values of (β_0, β_1) as unlikely to occur (in an intuitive sense).

Figure 3.3 illustrates the advantage of using a joint confidence region rather than individual confidence intervals. The region has a smaller area than the rectangle obtained from the individual intervals. However, as we want to find a good experimental design, we might wonder: *can we get a better confidence region (one with smaller area) than the one we have?* The answer is "yes!"

We need to use the following fact.

Result 3.4.1. *Let $\boldsymbol{x}$ and $\boldsymbol{x}_0$ be $p \times 1$ vectors, and let $\boldsymbol{S}$ be a symmetric positive definite $p \times p$ matrix. The volume of the hyperellipsoid*

$$(\boldsymbol{x} - \boldsymbol{x}_0)^\top \boldsymbol{S}^{-1} (\boldsymbol{x} - \boldsymbol{x}_0) = r^2$$

is given by

$$V = V_p \left[\det(\boldsymbol{S})\right]^{1/2} r^p, \tag{3.3}$$

where V_p is the volume of a p-dimensional unit hypersphere.

There are several aspects of Result 3.4.1 that need not concern us. The first is the value of V_p. There is a formula for this, but all that matters here is that its value is a constant for a given value of p. (This value is the area of a circle of radius 1 in $p = 2$ dimensions, the volume of a sphere of radius 1 in $p = 3$ dimensions, ...) Secondly, the use of r^2 on the right-hand side (RHS) implies that the RHS must be positive. As our application of the formula has the RHS equal to $2 \times \text{ResMS} \times F_{p,n-p;1-\alpha}$, the RHS will be positive. We shall replace $\boldsymbol{x}$ by $\boldsymbol{\beta}$, $\boldsymbol{x}_0$ by $\hat{\boldsymbol{\beta}}$ and $\boldsymbol{S}^{-1}$ by $\boldsymbol{F}^\top \text{diag}(n_1, \ldots, n_s)\boldsymbol{F}$, giving $\boldsymbol{S} = [\boldsymbol{F}^\top \text{diag}(n_1, \ldots, n_s)\boldsymbol{F}]^{-1}$.

In seeking to minimise the volume of the hyperellipsoid in (3.3), the only quantity that we can alter by selection of the design is $\left[\det(\boldsymbol{S})\right]^{1/2} = \left\{\det[\boldsymbol{F}^\top \text{diag}(n_1, \ldots, n_s)\boldsymbol{F}]^{-1}\right\}^{1/2} = \left\{\det\left[\boldsymbol{F}^\top \text{diag}(n_1, \ldots, n_s)\boldsymbol{F}\right]\right\}^{-1/2}$. The minimisation of this determinant is equivalent to the maximisation of $\det\left[\boldsymbol{F}^\top \text{diag}(n_1, \ldots, n_s)\boldsymbol{F}\right]$.

For the example of simple linear regression with $N = 10$ points and $p = 2$ parameters, we have

$$\boldsymbol{F}^\top \text{diag}(n_1, \ldots, n_s)\boldsymbol{F} = \begin{bmatrix} 10 & \sum_{i=1}^{10} x_i \\ \sum_{i=1}^{10} x_i & \sum_{i=1}^{10} x_i^2 \end{bmatrix}.$$

So

$$\det\left[\boldsymbol{F}^\top \text{diag}(n_1, \ldots, n_s)\boldsymbol{F}\right] = 10\sum_{i=1}^{10} x_i^2 - (\sum_{i=1}^{10} x_i)^2 = 10\sum_{i=1}^{10}(x_i - \bar{x})^2,$$

and we have already seen that this is maximised by maximising the spread of the x-values: setting five of the x-values equal to 0 and five of them equal to 1. This is the "extremities" design. It follows (see Example 1.5.1) that

$$\boldsymbol{F}^\top \text{diag}(n_1, \ldots, n_s)\boldsymbol{F} = \begin{bmatrix} 10 & 5 \\ 5 & 5 \end{bmatrix}.$$

You might be wondering whether there is really much difference between the confidence regions generated by our two candidate designs. Figure 3.4 illustrates the 95% confidence regions obtained for the two designs using the earlier values of $\hat{\beta}_0 = 0.74$, $\hat{\beta}_1 = 1.28$ and ResMS $= 4.35$. (These are kept constant

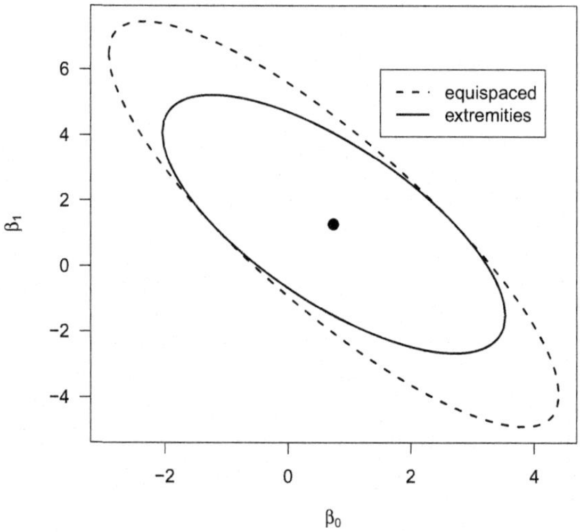

Figure 3.4 *95% confidence regions for the "equispaced" and "extremities" designs of Example 1.5.1. The point in the centre is* $(\hat{\beta}_0, \hat{\beta}_1)$. *The "extremities" design gives an ellipse of lesser area.*

so that only the design has changed.) No extra observations have been taken — the only difference between the designs is the location of the x-values at which the observations were made. But the confidence region for the optimal ("extremities") design has much smaller area than the region for the alternative ("equispaced") design. Choosing the optimal design was very worthwhile.

3.4.2 The standardised variance

In straight line regression, the line $\hat{y} = \hat{\beta}_0 + \hat{\beta}_1 x$ is used to predict the value of y for a specified value of x. Irrespective of the design, the expected value of $\hat{\beta}_0 + \hat{\beta}_1 x$ is $\beta_0 + \beta_1 x$. However, the variability of estimates around the true value depends upon the design that was chosen. So we might choose between competing designs on the basis of minimising the variance of $\hat{\beta}_0 + \hat{\beta}_1 x$. This variance is a function of x, so we could calculate $\mathrm{var}(\hat{\beta}_0 + \hat{\beta}_1 x)$ for all values of x in the design space, and choose the design that gives the most desirable values of the variance.

Note that $\hat{\beta}_0 + \hat{\beta}_1 x = \boldsymbol{f}^{\top}(x)\hat{\boldsymbol{\beta}}$, where $\boldsymbol{f}^{\top}(x) = (1, x)$. Then

$$\mathrm{var}(\hat{\beta}_0 + \hat{\beta}_1 x) = \mathrm{var}\left[\boldsymbol{f}^{\top}(x)\hat{\boldsymbol{\beta}}\right] = \boldsymbol{f}^{\top}(x)\mathrm{cov}(\hat{\boldsymbol{\beta}})\boldsymbol{f}(x).$$

From (3.2), it follows that

$$\text{var}(\hat{\beta}_0 + \hat{\beta}_1 x) = \sigma^2 \boldsymbol{f}^\top(x)[\boldsymbol{F}^\top \text{diag}(n_1, \ldots, n_s)\boldsymbol{F}]^{-1}\boldsymbol{f}(x).$$

A disadvantage of basing a comparison between designs on this variance is that the variance depends directly on the number of observations being taken on a design. If the number of observations made at each support point is doubled, the variance will be halved. In order to ensure that a comparison of designs is "fair", we need to take account of the value of N, the total number of observations used in a design. Similarly, we also need to assume that the error variance, σ^2, is the same for the two designs. These two aims can be met by considering not $\text{var}(\hat{\beta}_0 + \hat{\beta}_1 x)$, but instead $N \times \text{var}(\hat{\beta}_0 + \hat{\beta}_1 x)/\sigma^2$, which is called the *standardised variance.* That is,

$$\text{standardised variance} = N \boldsymbol{f}^\top(x)[\boldsymbol{F}^\top \text{diag}(n_1, \ldots, n_s)\boldsymbol{F}]^{-1}\boldsymbol{f}(x). \qquad (3.4)$$

This allows a comparison of different designs, irrespective of the values of N and σ^2.

Although the formula for the standardised variance was derived for a standard straight line regression, it is in fact also true for a quadratic, cubic, ... polynomial of x.

Example 3.4.2. *Example 1.5.1 considered two designs of $N = 10$ observations each: the "equispaced" and "extremities" designs. The designs were intended for the model $Y_i = \beta_0 + \beta_1 x_i + E_i$ $(i = 1, \ldots, 10)$ for values of x between 0 and 1. Using the values of $[\boldsymbol{F}^\top \text{diag}(n_1, \ldots, n_s)\boldsymbol{F}]^{-1}$ for the two designs, as found in Example 1.5.1, the standardised variances for the two designs have been calculated for all x between 0 and 1. The standardised variances are plotted in Figure 3.5.*

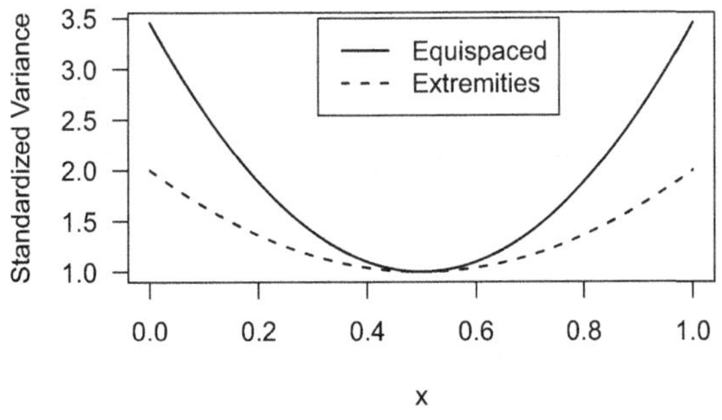

Figure 3.5 *Plots of the standardised variances of the "equispaced" and "extremities" designs for all x satisfying $0 \leq x \leq 1$.*

It is evident from Figure 3.5 that the standardised variance for the "extremities" design is less than the standardised variance for the "equispaced" design for all values of x except 0.5, where the two standardised variances are equal. As the aim is to choose the design with the smaller standardised variance, we would prefer the "extremities" design.

3.5 Generalized linear models

3.5.1 Theory

Recall from (1.21) that the covariance matrix of the estimated parameter vector, $\hat{\boldsymbol{\beta}}$, is $\boldsymbol{\mathcal{I}}^{-1}$, where the (j,k) element of $\boldsymbol{\mathcal{I}}$ is

$$\mathcal{I}_{jk} = \sum_{i=1}^{s} n_i \frac{f_{ij} f_{ik}}{\text{var}(Y_i)} \left(\frac{\partial \mu_i}{\partial \eta_i}\right)^2, \quad j,k \in \{0,\ldots,p-1\}.$$

As $f_{ij} f_{ik}$ is the (j,k) element of $\boldsymbol{f}_i \boldsymbol{f}_i^{\top}$, we can write

$$\boldsymbol{\mathcal{I}} = \sum_{i=1}^{s} n_i \frac{1}{\text{var}(Y_i)} \left(\frac{\partial \mu_i}{\partial \eta_i}\right)^2 \boldsymbol{f}_i \boldsymbol{f}_i^{\top}.$$

Recall from (1.12) that it is usual to write var(Y) as $\phi V(\mu)$. So $1/\text{var}(Y_i)$ may be replaced by $1/[\phi V(\mu_i)]$. It will be convenient to write

$$\omega_i = \omega(\boldsymbol{x}_i) = \frac{1}{\phi V(\mu_i)} \left(\frac{\partial \mu_i}{\partial \eta_i}\right)^2, \quad i = 1,\ldots,s, \tag{3.5}$$

and to call $\omega(\boldsymbol{x}_i)$ the *model weight* at the ith support point. Note that $\omega(\boldsymbol{x}_i)$ is actually a function of $\text{var}(Y_i) = \phi V(\mu_i)$, and of the link function $g(\cdot)$, $\boldsymbol{\beta}$ and $\boldsymbol{x}_i$ through the relationships $g(\mu_i) = \eta_i = \boldsymbol{f}^{\top}(\boldsymbol{x}_i)\boldsymbol{\beta}$ and $\text{var}(Y) = \phi V(\mu)$. However, for a given design problem, the functions $V(\mu)$, $g(\mu)$ and the quantities ϕ and $\boldsymbol{\beta}$ will not vary from support point to support point, so we write the design weight as $\omega(\boldsymbol{x}_i)$ for simplicity.

In the discussion of the standardised variance on page 61, we saw the advantage of removing the effect of the total number of observations, N, from consideration of an exact design. So the expression for $\boldsymbol{\mathcal{I}}$ will be divided by N to achieve this, giving the matrix

$$\frac{1}{N}\sum_{i=1}^{s} n_i\, \omega(\boldsymbol{x}_i) \boldsymbol{f}(\boldsymbol{x}_i) \boldsymbol{f}^{\top}(\boldsymbol{x}_i) = \sum_{i=1}^{s} \delta_i\, \omega(\boldsymbol{x}_i) \boldsymbol{f}(\boldsymbol{x}_i) \boldsymbol{f}^{\top}(\boldsymbol{x}_i),$$

where $\delta_i = n_i/N$ is the design weight of the ith support point (see page 50). This is written as

$$\boldsymbol{M}(\xi, \boldsymbol{\beta}) = \sum_{i=1}^{s} \delta_i\, \omega(\boldsymbol{x}_i) \boldsymbol{f}(\boldsymbol{x}_i) \boldsymbol{f}^{\top}(\boldsymbol{x}_i). \tag{3.6}$$

The matrix $\boldsymbol{M}(\xi, \boldsymbol{\beta})$ is called the *information matrix*. It is a function of the design ξ through the design weights and values of the support points, and it is a function of $\boldsymbol{\beta}$ through the relationship $g(\mu) = \boldsymbol{f}^\top(\boldsymbol{x})\boldsymbol{\beta}$ and $\mathrm{var}(Y) = \phi V(\mu)$.

While $\boldsymbol{M}(\xi, \boldsymbol{\beta})$ was introduced for exact designs, it is convenient to use exactly the same definition in (3.6) for approximate designs (see page 50), where the design weight δ_i represents an idealised weight, or proportion of observations, at the ith support point.

Notation: Some writers on designs for GLMs describe both δ_i and ω_i simply as "weights." In this book, to avoid confusion, the quantities δ_i will always be described as design weights, while the quantities $\omega_i(\boldsymbol{x}_i)$ will be called model weights. A design is characterised by the values of δ_i and $\boldsymbol{x}_i$ that are selected, but the model weights for that design may vary according to the model that is chosen, which depends on the link function $g(\mu)$, the distribution of the observations, and the vector, $\boldsymbol{f}(\boldsymbol{x})$, that is used in the linear predictor $\eta = \boldsymbol{f}^\top(\boldsymbol{x})\boldsymbol{\beta}$.

3.5.2 *Example: a simple logistic regression*

Consider a logistic regression. Suppose that it is planned to conduct an experiment with a binary outcome, $Y \in \{0, 1\}$, and we wish to determine the relationship between the probability, π, of obtaining a 1 ("success") and the value of a single explanatory variable, x. We take n_i observations at $x = x_i$ $(i = 1, \ldots, s)$, with $\sum_i n_i = N$. Denote by y_{ij} the value of the jth observation at x_i $(j = 1, \ldots, n_i;\ i = 1, \ldots, s)$.

Further suppose that the logit link function is used to predict π in terms of x, with just $p = 2$ parameters; i.e., $\ln[\pi/(1-\pi)] = \eta = \beta_0 + \beta_1 x$. Write $\eta_i = \beta_0 + \beta_1 x_i$ $(i = 1, \ldots, s)$. This corresponds to

$$\pi_i = \frac{\exp(\eta_i)}{1 + \exp(\eta_i)} \times \frac{\exp(-\eta_i)}{\exp(-\eta_i)} = \frac{1}{\exp(-\eta_i) + 1}. \tag{3.7}$$

We can calculate $(\partial \pi_i)/(\partial \eta_i)$ directly from (3.7), giving

$$\frac{\partial \pi_i}{\partial \eta_i} = \frac{\exp(-\eta_i)}{[\exp(-\eta_i) + 1]^2} = \frac{1}{\exp(-\eta_i) + 1} \times \frac{\exp(-\eta_i)}{\exp(-\eta_i) + 1} = \pi_i(1 - \pi_i).$$

Alternatively, we can start from $\eta_i = \ln[\pi_i/(1 - \pi_i)]$, and get

$$\frac{\partial \eta_i}{\partial \pi_i} = \frac{1}{\pi_i(1 - \pi_i)}.$$

Both approaches lead to the result

$$\frac{\partial \pi_i}{\partial \eta_i} = \pi_i(1 - \pi_i). \tag{3.8}$$

A standard result for the Bernoulli distribution (or a binomial(n, π) distribution with $n = 1$) is that $\mathrm{var}(Y_{ij}) = \pi_i(1 - \pi_i)$. Given the equation

$\eta_i = \beta_0 + \beta_1 x_i = \boldsymbol{f}(x_i)^\top \boldsymbol{\beta}$, then $f_0(x_i) = 1$ and $f_1(x_i) = x_i$. So (1.19) and (1.21) lead to the score statistics

$$\begin{aligned} U_0 &= \sum_{i=1}^{s}\sum_{j=1}^{n_i} \frac{(y_{ij} - \pi_i)}{\pi_i(1-\pi_i)} \times 1 \times \pi_i(1-\pi_i) \\ &= \sum_{i=1}^{s}\sum_{j=1}^{n_i} (y_{ij} - \pi_i) = \sum_{i=1}^{s} n_i(\bar{y}_{i\cdot} - \pi_i), \end{aligned} \tag{3.9}$$

and

$$\begin{aligned} U_1 &= \sum_{i=1}^{s}\sum_{j=1}^{n_i} \frac{(y_{ij} - \pi_i)}{\pi_i(1-\pi_i)} \times x_i \times \pi_i(1-\pi_i) \\ &= \sum_{i=1}^{s}\sum_{j=1}^{n_i} (y_{ij} - \pi_i)x_i = \sum_{i=1}^{s} n_i(\bar{y}_{i\cdot} - \pi_i)x_i, \end{aligned} \tag{3.10}$$

and to the values of $I_{jk} = \text{cov}(U_j, U_k)$:

$$\begin{aligned} I_{00} &= \sum_{i=1}^{s}\sum_{j=1}^{n_i} \frac{1 \times 1}{\pi_i(1-\pi_i)} \times [\pi_i(1-\pi_i)]^2 \\ &= \sum_{i=1}^{s} n_i \pi_i(1-\pi_i), \\ I_{01} = I_{10} &= \sum_{i=1}^{s}\sum_{j=1}^{n_i} \frac{1 \times x_i}{\pi_i(1-\pi_i)} \times [\pi_i(1-\pi_i)]^2 \\ &= \sum_{i=1}^{s} n_i \pi_i(1-\pi_i)x_i, \end{aligned}$$

and

$$\begin{aligned} I_{11} &= \sum_{i=1}^{s}\sum_{j=1}^{n_i} \frac{x_i \times x_i}{\pi_i(1-\pi_i)} \times [\pi_i(1-\pi_i)]^2 \\ &= \sum_{i=1}^{s} n_i \pi_i(1-\pi_i)x_i^2. \end{aligned}$$

Let $\hat{\pi}_i = 1/[\exp(-\hat{\beta}_0 - \hat{\beta}_1 x_i) + 1]$ $(i = 1, \ldots, s)$, where $\hat{\beta}_0$ and $\hat{\beta}_1$ are the ML estimators of β_0 and β_1. It follows from (1.17), (3.9) and (3.10) that the ML estimators satisfy

$$\begin{aligned} \sum_{i=1}^{s} n_i(\bar{y}_{i\cdot} - \hat{\pi}_i) &= 0, \\ \sum_{i=1}^{s} n_i(\bar{y}_{i\cdot} - \hat{\pi}_i)x_i &= 0. \end{aligned} \tag{3.11}$$

For $s = 2$, (3.11) becomes

$$\begin{aligned} n_1(\bar{y}_{1\cdot} - \hat{\pi}_1) + n_2(\bar{y}_{2\cdot} - \hat{\pi}_2) &= 0 \\ n_1(\bar{y}_{1\cdot} - \hat{\pi}_1)x_1 + n_2(\bar{y}_{2\cdot} - \hat{\pi}_2)x_2 &= 0, \end{aligned}$$

and a solution is

$$\hat{\pi}_1 = 1/[\exp(-\hat{\beta}_0 - \hat{\beta}_1 x_1) + 1] = \bar{y}_{1\cdot}, \tag{3.12}$$

$$\hat{\pi}_2 = 1/[\exp(-\hat{\beta}_0 - \hat{\beta}_1 x_2) + 1] = \bar{y}_{2\cdot}. \tag{3.13}$$

For $x_1 \neq x_2$, these are two linearly independent equations in two unknowns ($\hat{\beta}_1$ and $\hat{\beta}_2$), so the solution will be unique. Equations (3.12) and (3.13) simplify to

$$\exp(-\hat{\beta}_0 - \hat{\beta}_1 x_1) = \frac{1}{\bar{y}_{1\cdot}} - 1 \quad \text{and} \quad \exp(-\hat{\beta}_0 - \hat{\beta}_1 x_2) = \frac{1}{\bar{y}_{2\cdot}} - 1,$$

which reduce to

$$\begin{aligned} \hat{\beta}_0 + \hat{\beta}_1 x_1 = \ln[\bar{y}_{1\cdot}/(1 - \bar{y}_{1\cdot})] &= \text{logit}(\bar{y}_{1\cdot}) \\ \hat{\beta}_0 + \hat{\beta}_1 x_2 = \ln[\bar{y}_{2\cdot}/(1 - \bar{y}_{2\cdot})] &= \text{logit}(\bar{y}_{2\cdot}). \end{aligned}$$

These equations imply

$$\begin{aligned} \hat{\beta}_1 &= \frac{\text{logit}(\bar{y}_{1\cdot}) - \text{logit}(\bar{y}_{2\cdot})}{x_1 - x_2} \\ \hat{\beta}_0 &= \frac{x_1 \text{logit}(\bar{y}_{2\cdot}) - x_2 \text{logit}(\bar{y}_{1\cdot})}{x_1 - x_2}, \end{aligned}$$

or equivalently

$$\hat{\beta}_1 = \frac{\ln[y_{1\cdot}/(n_1 - y_{1\cdot})] - \ln[y_{2\cdot}/(n_2 - y_{2\cdot})]}{x_1 - x_2} \tag{3.14}$$

$$\hat{\beta}_0 = \frac{x_1 \ln[y_{2\cdot}/(n_2 - y_{2\cdot})] - x_2 \ln[y_{1\cdot}/(n_1 - y_{1\cdot})]}{x_1 - x_2}. \tag{3.15}$$

Recall that $y_{i\cdot}$ represents the sum of all y_{ij} for $j = 1, \ldots, n_i$ $(i = 1, 2)$.

For $s > 2$, the equations that are analogous to (3.12) and (3.13) are

$$\hat{\pi}_i = 1/[\exp(-\hat{\beta}_0 - \hat{\beta}_1 x_i) + 1] = \bar{y}_{i\cdot} \quad (i = 1, \ldots, s).$$

These s equations do not generally form a solution to (3.11), because the solution in (3.14) and (3.15) from $i = 1$ and $i = 2$ cannot be expected to solve the additional equations $\hat{\pi}_i = 1/[\exp(-\hat{\beta}_0 - \hat{\beta}_1 x_i)] = \bar{y}_{i\cdot}$ for $i > 2$. Instead, an iterative solution to (3.11) must be found. The method of obtaining an iterative solution is described in Dobson & Barnett (2008, p. 66), but is not of direct relevance to a consideration of designing the experiment.

Example 3.5.1. *Consider an experiment with* $s = 3$ *support points, where* $n_1 = n_2 = n_3 = 5$, $x_1 = 0$, $x_2 = 0.5$ *and* $x_3 = 1$. *Suppose that analysing the experiment leads to* $\bar{y}_1 = 1/5$, $\bar{y}_2 = 2/5$ *and* $\bar{y}_3 = 2/5$. *A logistic regression using the glm function in R gives the estimates* $\hat{\beta}_0 = -1.1752$ *and* $\hat{\beta}_1 = 0.9174$. *This leads to the estimated probabilities of success at the three support points:* $\hat{\pi}_1 = 1/[\exp(1.1752 - 0.9174 \times x_1) + 1] = 0.2359$, $\hat{\pi}_2 = 0.3282$ *and* $\hat{\pi}_3 = 0.4439$. *Substituting these values into the left-hand sides of the two equations in* (3.11) *gives 0 (to within rounding error) in each case. This suggests that* $\hat{\beta}_0 = -1.1752$ *and* $\hat{\beta}_1 = 0.9174$ *are the ML estimates of* β_0 *and* β_1 *although, strictly speaking, one should also check that the matrix of partial second derivatives of* ℓ *is negative definite; see the lines immediately below* (1.17).

3.6 Difficulties caused by small samples

Consider again the solution in (3.14) and (3.15) for a logistic regression with $s = 2$ support points. For $n_1 = n_2 = 5$, $x_1 = 0$, $x_2 = 1$ and $\bar{y}_{1\cdot} = 3/5$ and $\bar{y}_{2\cdot} = 2/5$, the statistical packages R, SAS and SPSS all give (to varying numbers of decimal places) the same answer as (3.14) and (3.15): $\hat{\beta}_0 = 0.4054651$ and $\hat{\beta}_1 = -0.8109302$. The result from the software agrees with the theory, as one would expect.

However, as each y_{ij} is either 0 or 1, then $0 \leq \bar{y}_{i\cdot} \leq 1$ for each i, and $\text{logit}(\bar{y}_{i\cdot})$ will be undefined if the y_{ij} ($j = 1, \ldots, n_i$) are all 0, or all 1. In this case, $\hat{\beta}_0$ and $\hat{\beta}_1$ should be undefined. You might hope that this would produce an error message from your software, but this does not necessarily occur. For example, for $n_1 = n_2 = 5$, $x_1 = 0$, $x_2 = 1$, $\bar{y}_{1\cdot} = 0/5$ and $\bar{y}_{2\cdot} = 3/5$, R gives the solution

```
               Estimate Std. Error         z value  Pr(>|z|)
(Intercept) -24.53650   57729.93 -0.0004250222 0.9996609
x            24.94197   57729.93  0.0004320457 0.9996553
```

SAS's *proc genmod* gives estimates of $\hat{\beta}_0 = -27.3633$ and $\hat{\beta}_1 = 27.7708$. The SPSS *generalized linear models* command gives the message, "A quasi-complete separation may exist in the data. The maximum likelihood estimates do not exist," but it still provides a solution of $\hat{\beta}_0 = -22.566$ and $\hat{\beta}_1 = 22.972$. To be fair, a researcher who looks at the standard errors of the estimates in the output from any of these three packages would realise that something is wrong because the values of the standard errors are very large. However, if you are simply performing simulations of an experiment and then using the software to evaluate the estimates, you would be seriously misled.

The problem of infinite values for the estimates $\hat{\beta}_0, \ldots, \hat{\beta}_{p-1}$ also occurs for $s > 2$ support points. Its cause cannot be so easily demonstrated, as explicit expressions for the estimates do not exist when $s > p$. However, various authors have investigated the matter. Albert & Atkinson (1984) considered the set of n observations $(y_i, \boldsymbol{x}_i)$ from the experiment, and classified them into three mutually exclusive and exhaustive categories: 'complete separation,' 'quasi-complete separation' and 'overlap.' Only if the observations belong to the 'overlap' category are the ML estimates finite.

The UCLA Statistical Consulting Group has a nice article (UCLA: Statistical Consulting Group, 2015) giving examples of both complete separation and quasi-complete separation, and showing examples of error messages that it was able to obtain from the packages SAS, SPSS, and Stata. In the absence of such error messages, it seems best to look for large values of the standard errors. If a standard error is greater than 10, it is certainly worth having a human (and not a computer) look at the output to see if a separation problem is likely to have occurred.

As separation is a data analysis issue (i.e., it does not arise until after the data have been collected), you may wonder why it is being considered in a book on design. The first reason is that a good designer will not consider a design completely in isolation of the subsequent analysis: if the data from the design cannot be analysed in some particular circumstance, then the design is of little value. Secondly, a good designer should try to mitigate the circumstances which prevent an analysis. In this case, the problem is the occurrence of separation, and we know that larger samples will reduce the risk of the occurrence of this problem.

For logistic regression with the simple model $\eta = \beta_0 + \beta_1 x$, it is straightforward to calculate the probability that separation occurs when $s = 2$. Write $\pi_i = 1/[1+\exp(-\eta_i)]$ $(i = 1, 2)$ for the probability of a "success" occurring when $x = x_i$, and let n_i be the number of observations made at x_i. Then the probability that Y_i equals 0 or n_i is

$$p_i = \binom{n_i}{0}\pi_i^0(1-\pi_i)^{n_i} + \binom{n_i}{n_i}\pi_i^{n_i}(1-\pi_i)^0 = \pi_i^{n_i} + (1-\pi_i)^{n_i},$$

and so the probability that Y_i is neither 0 nor n_i is $(1-p_i)$. The values of Y_1 and Y_2 are independent of one another, so the probability that separation does *not* occur (neither Y_1 nor Y_2 take their minimum or maximum possible values) is $(1-p_1)\times(1-p_2)$. Therefore the probability that separation *does* occur is

$$\begin{aligned} p_{\text{sep}} &= 1-(1-p_1)(1-p_2) \\ &= 1-\{1-[\pi_1^{n_1}+(1-\pi_1)^{n_1}]\}\{1-[\pi_2^{n_2}+(1-\pi_2)^{n_2}]\}. \qquad (3.16) \end{aligned}$$

Example 3.6.1. *Assume that $\beta_0 = 0$ and $\beta_1 = 1$, and choose $x_0 = 0$, $x_1 = 1$ and $n_1 = n_2 = 4$. Then $\eta_i = \beta_0 + \beta_1 x_i = x_i$, and $\pi_i = 1/[1+\exp(-x_i)]$. This gives $\pi_1 = 0.5$ and $\pi_2 = 0.7311$. So, from (3.16), the probability that an experiment with these values of n_1, n_2, x_1 and x_2 will yield a result that produces separation is*

$$1-[1-(0.5^4+0.5^4)][1-(0.7311^4+0.2689^4)] = 0.3796.$$

If n_1 and n_2 are each increased to 10, the probability decreases to

$$1-[1-(0.5^{10}+0.5^{10})][1-(0.7311^{10}+0.2689^{10})] = 0.0455,$$

while if $n_1 = n_2 = 20$, the probability of separation occurring is 0.0019.

Clearly the probability that the observations will not overlap decreases as each value of n_i $(i = 1, \ldots, s)$ increases. For a simple situation such as this one, it is very easy to calculate in advance of the experiment the risk that separation occurs, and it would be unwise to conduct the experiment with the planned values of n_1 and n_2 if the risk of separation occurring is unacceptably high.

However, as the number of support points increases, it becomes more difficult to calculate the probability of separation occurring. For example, for the same values of β_0 and β_1, three support points at 0, 0.5 and 1, and $n_1 = n_2 = n_3 = 4$, you might expect the results $Y_1 = 0$, $Y_2 = 4$ and $Y_3 = 1$ to lead to separation, but they do not. You can use simulation to estimate the probability of separation occurring. It is straightforward to simulate the conduct of a designed experiment for various values of $n_1, \ldots, n_s$ and nominated values of the parameters $\beta_0, \ldots, \beta_{p-1}$, although obviously the running time increases rapidly as the numbers of predictor variables and support points increase. Program_8 in the online repository is set up to investigate s = 3 points for $p = 2$ parameters. I have specified values of β_0 and β_1, and x_1, x_2 and x_3 for illustrative purposes only: no claim is made for any optimality of the design.

For $\beta_0 = 0$ and $\beta_1 = 1$, $x_1 = 0$, $x_2 = 0.5$ and $x_3 = 1$, and a seed of 12345, 1000 simulations led to a proportion of 0.072 of data sets for which at least one of s.e.$(\hat{\beta}_0)$ or s.e.$(\hat{\beta}_1)$ exceeded 10 when $n_1 = n_2 = n_3 = 4$. (For each of the relevant simulations, the warning `glm.fit: fitted probabilities numerically 0 or 1 occurred` was given.) Upon increasing the sample sizes to $n_1 = n_2 = n_3 = 10$ (and changing the seed to 54321), the proportion of samples with at least one standard error greater than 1 was 0.

When $x_1 = -1$, $x_2 = 0$ and $x_3 = 0$ were used as the support points, sample sizes of $n_1 = n_2 = n_3 = 4$ gave the result that the proportion of samples with at least one standard error greater than 1 was 0.111; when $n_1 = n_2 = n_3 = 10$ were used, the proportion of samples with at least one standard error greater than 1 was 0.001.

There are two points to be drawn from this investigation:

- A sample that contains all zeros or all ones is more likely to occur if n_i is small. If you must use small sample sizes, do not use the standard maximum likelihood method of estimation. An alternative method (and how to design an experiment for it) will be described in Section 4.8.
- Any analysis (or a simulated investigation of the properties of the estimators) should not consider only the values of the estimates. You must consider the values of the standard errors as well.

3.7 Optimality

3.7.1 Number of support points

In seeking an optimal design, an important question is *how many support points are needed?*

The minimum number is p, the number of parameters. This is intuitively obvious. At least two separate points are needed to estimate the two parameters, β_0 and β_1, of a straight line, and at least three separate points are required to estimate the three parameters, β_0, β_1 and β_2, of a quadratic. See Figure 3.2 to recall the situation that arises when there are only two points for a quadratic.

The maximum number of support points necessary for an optimal design can be shown to be $p(p+1)/2+1$, using *Carathéodory's Theorem.* The proof is beyond the scope of this book, but is given in Rockafellar (1970, p. 155) or Silvey (1980, p. 77). Pukelsheim (1993, p. 190) showed that, if our interest is in estimating all elements of the parameter vector $\boldsymbol{\beta}$, then this upper limit can be reduced to $p(p+1)/2$.

Comments

1. If the response variable is a real number, the limits of p and $p(p+1)/2$ apply without exception.

2. If the response variable is categorical, the limits of p and $p(p+1)/2$ are for experiments where the response variable has only two categories as its possible values. A Bernoulli random variable is an example of this. It has two possible values, "success" and "failure," but by recording the number of successes (out of one), one automatically knows the number of failures.

   In contrast, a multinomial random variable that has $k\,(>2)$ possible values (e.g., an experimental animal that receives a drug may Die, Partially Recover or Fully Recover) requires the values of $(k-1)$ of the categories to be recorded before the last category is automatically known. Generalized linear models are available to predict the outcomes of $(k-1)$ of the categories from knowledge of the explanatory variable(s). In such situations, the values observed on some categories may provide information about the parameters being used to predict the values of other categories, and the limits of p and $p(p+1)/2$ may be able to be reduced. This is discussed further in Section 6.3.

   The limits of p and $p(p+1)/2$ for the number of support points may not be appropriate for Bayesian designs. See Chapter 7.

3. A design with only p support points uses all the points to estimate the parameters. There are no points available to check whether the model is appropriate; e.g., whether a quadratic model should have been fitted instead of a straight line. This will be considered in Sub-section 3.7.5.

Example 3.7.1. *When $p=2$, a design will have between $p=2$ and $p(p+1)/2=3$ support points. When p is 4, between $p=4$ and $p(p+1)/2=10$ support points will be required. If there are m explanatory variables, then a computational search over s possible support points will have $v=ms+s$ variables whose values are to be found: the m coordinates of each of the s support points, and the s design weights.*

When searching for an optimal design, I would start with $p(p+1)/2$ support points, and hope to find that the optimisation routine gives several design weights that are effectively 0. I would then reduce the number of support points in the next search.

3.7.2 *Optimality criteria*

To select a design that is "optimal," you need to decide what criteria the design must satisfy for you to consider it optimal. There are numerous criteria in common use, and a design that is optimal under some criteria need not necessarily be optimal under others. (If you have ever looked at two people in a relationship and wondered what one of them sees in the other, then the previous sentence should not surprise you at all.)

Most criteria come under the heading of "alphabet optimality" because the criterion's name is a letter from the alphabet (e.g., D-optimality). An extensive list of alphabetical optimality criteria are considered in Atkinson & Donev (1992, Chapter 10). That discussion relates to designs under the general linear model, and some minor adjustments would need to be made when considering criteria applying to GLMs.

Table 3.2 provides a list of five commonly used alphabet criteria. There are numerous other criteria that are used less frequently.

Optimality Criterion	*The criterion seeks a design that minimises the*
A-	average of the variances of the parameter estimates
D-	volume of a confidence hyperellipsoid for the parameters
D_S-	volume of a confidence hyperellipsoid for a subset of the parameters, taking into account the presence of the remaining parameters
E-	maximum variance of a linear combination of parameter estimates, $\boldsymbol{a}^\top\hat{\boldsymbol{\beta}}$, where $\boldsymbol{a}^\top\boldsymbol{a} = 1$
G-	maximum value of the standardised variance that occurs at any point in the design space

Table 3.2 *Common alphabet optimality criteria.*

3.7.3 *A-optimality*

A design ξ_A^* is A-optimal amongst a set of designs Ξ if it possesses the minimum value of the average of the variances of the parameter estimates.

For an experiment in which all the explanatory variables $x_1, \ldots, x_m$ are real variables, the A-optimality criterion is of little value. For example, if we have the linear model $\eta = \beta_0 + \beta_1 x_1 + \beta_2 x_2 + \beta_3 x_1 x_2$, then the average of the variances of the parameter estimates,

$$\frac{1}{4}\sum_{i=0}^{3} \mathrm{var}(\hat{\beta}_i),$$

has little meaning, and so it is unlikely that one would wish to choose a design that minimises this average.

However, if some of the variables are indicator variables that show which level of a treatment factor is associated with a particular observation (as in Examples 1.1.2 and 1.1.4), then it can be shown that the A-optimality criterion selects the design for which

$$\frac{1}{t(t-1)} \sum_{i=1}^{t} \sum_{\substack{j=1 \\ i \neq j}}^{t} \mathrm{var}(\hat{\tau}_i - \hat{\tau}_j)^2$$

is a minimum. That is, the A-optimal design minimises the average of the variances of the effects of the pairwise treatment differences, and this is indeed a sensible criterion to apply when we wish to compare the effects of the t different treatments.

As our interest in this book will mostly be with individual model parameters, rather than with comparisons of parameters, no further consideration will be given to the A-optimality criterion.

3.7.4 D-optimality

A design ξ_D^* is D-optimal amongst a set of designs Ξ if, for an arbitrary value of α, the design possesses the minimum volume of a $100(1-\alpha)\%$ confidence hyperellipsoid for the parameters in $\boldsymbol{\beta}$. As we saw on page 59 for the simple example of data from a normal distribution where the errors have a constant variance, minimising the volume of the hyperellipsoid is equivalent to maximising the determinant of the information matrix. In general, a D-optimal design, ξ_D^*, is that one amongst all $\xi \in \Xi$ for which $\det[M(\xi, \boldsymbol{\beta})]$ is maximised.

The D-optimality criterion receives most attention in this book.

It will sometimes be of interest to compare two designs, ξ_1 and ξ_2, from the perspective of D-optimality. As D-optimality seeks the design for which $\det[M(\xi, \boldsymbol{\beta})]$ is maximised, then one might examine the ratio $\det[M(\xi_1, \boldsymbol{\beta})]/\det[M(\xi_2, \boldsymbol{\beta})]$. Unfortunately, this simple suggestion has a disadvantage. It seems intuitive to regard ξ_1 as being twice as "good" as ξ_2 if $M(\xi_1, \boldsymbol{\beta}) = 2M(\xi_2, \boldsymbol{\beta})$, but unfortunately this implies that $\det[M(\xi_1, \boldsymbol{\beta})] = 2^p \det[M(\xi_2, \boldsymbol{\beta})]$; see the result on page 27. This would give a ratio of determinants of 2^p, rather than the intuitive 2. To avoid this counterintuitive event, the D-efficiency of ξ_1 relative to ξ_2 is given by

$$\text{D-efficiency of } \xi_1 \text{ relative to } \xi_2 = \left\{ \frac{\det[M(\xi_1, \boldsymbol{\beta})]}{\det[M(\xi_2, \boldsymbol{\beta})]} \right\}^{1/p}. \tag{3.17}$$

For D-optimality, there is a special result about the design weights of the support points when s takes the minimum value of p. In this case, the p points have equal design weights, namely $1/p$. See Silvey (1980, p.42).

3.7.5 D_s-optimality

Occasions exist when, although the expression for the linear predictor η has p parameters, our chief interest is in a subset of p_1 ($< p$) of them. We may write

$$\eta = \boldsymbol{f}(\boldsymbol{x})^\top \boldsymbol{\beta} = \boldsymbol{f}_1(\boldsymbol{x})^\top \boldsymbol{\beta}_1 + \boldsymbol{f}_2(\boldsymbol{x})^\top \boldsymbol{\beta}_2, \tag{3.18}$$

where $\boldsymbol{\beta}_1$ is $p_1 \times 1$ and contains the p_1 parameters of special interest, while $\boldsymbol{\beta}_2$ is $p_2 \times 1$ and contains the remaining $p_2 = p - p_1$ parameters of the full model.

Example 3.7.2. *Suppose that we are considering the quadratic model* $\eta = \beta_0 + \beta_1 x_1 + \beta_2 x_2 + \beta_3 x_1^2 + \beta_4 x_1 x_2 + \beta_5 x_2^2$, *but are particularly interested in estimating* β_3, β_4 *and* β_5. *Then the full model has* $m = 2$ *mathematically independent explanatory variables and* $p = 6$ *parameters, and* (3.18) *follows with* $p_1 = p_2 = 3$, $\boldsymbol{\beta}_1 = (\beta_3, \beta_4, \beta_5)^\top$, $\boldsymbol{f}_1(\boldsymbol{x}) = (x_1^2, x_1 x_2, x_2^2)^\top$, $\boldsymbol{\beta}_2 = (\beta_0, \beta_1, \beta_2)^\top$ *and* $\boldsymbol{f}_2(\boldsymbol{x}) = (1, x_1, x_2)^\top$.

For the general situation, if necessary rearrange the order of the elements of $\boldsymbol{\beta}$ and $\boldsymbol{f}(\boldsymbol{x})$ so that

$$\boldsymbol{\beta} = \begin{bmatrix} \boldsymbol{\beta}_1 \\ \boldsymbol{\beta}_2 \end{bmatrix} \quad \text{and} \quad \boldsymbol{f}(\boldsymbol{x}) = \begin{bmatrix} \boldsymbol{f}_1(\boldsymbol{x}) \\ \boldsymbol{f}_2(\boldsymbol{x}) \end{bmatrix}.$$

For a design ξ as given in (3.1) and appropriate model weights $\omega(\boldsymbol{x}_i)$ ($i = 1, \ldots, s$), the information matrix specified in (3.6) is

$$\begin{aligned} \boldsymbol{M}(\xi, \boldsymbol{\beta}) &= \sum_{i=1}^{s} \delta_i\, \omega(\boldsymbol{x}_i) \boldsymbol{f}(\boldsymbol{x}_i) \boldsymbol{f}^\top(\boldsymbol{x}_i) \\ &= \sum_{i=1}^{s} \delta_i\, \omega(\boldsymbol{x}_i) \begin{bmatrix} \boldsymbol{f}_1(\boldsymbol{x}) \\ \boldsymbol{f}_2(\boldsymbol{x}) \end{bmatrix} \left[\boldsymbol{f}_1^\top(\boldsymbol{x}), \boldsymbol{f}_2^\top(\boldsymbol{x}) \right] \\ &= \begin{matrix} & p_1 & p_2 \\ \begin{matrix} p_1 \\ p_2 \end{matrix} & \multicolumn{2}{c}{\begin{bmatrix} \boldsymbol{M}_{11}(\xi, \boldsymbol{\beta}) & \boldsymbol{M}_{12}(\xi, \boldsymbol{\beta}) \\ \boldsymbol{M}_{21}(\xi, \boldsymbol{\beta}) & \boldsymbol{M}_{22}(\xi, \boldsymbol{\beta}) \end{bmatrix}} \end{matrix}, \end{aligned}$$

where

$$\boldsymbol{M}_{jk}(\xi, \boldsymbol{\beta}) = \sum_{i=1}^{s} \delta_i\, \omega(\boldsymbol{x}_i) \boldsymbol{f}_j(\boldsymbol{x}_i) \boldsymbol{f}_k^\top(\boldsymbol{x}_i) \quad (j, k \in \{1, 2\}).$$

Let $\boldsymbol{B}(\boldsymbol{\xi}, \boldsymbol{\beta}) = \boldsymbol{M}^{-1}(\xi, \boldsymbol{\beta})$, and partition $\boldsymbol{B}(\boldsymbol{\xi}, \boldsymbol{\beta})$ as

$$\boldsymbol{B}(\boldsymbol{\xi}, \boldsymbol{\beta}) = \begin{matrix} & p_1 & p_2 \\ \begin{matrix} p_1 \\ p_2 \end{matrix} & \multicolumn{2}{c}{\begin{bmatrix} \boldsymbol{B}_{11}(\xi, \boldsymbol{\beta}) & \boldsymbol{B}_{12}(\xi, \boldsymbol{\beta}) \\ \boldsymbol{B}_{21}(\xi, \boldsymbol{\beta}) & \boldsymbol{B}_{22}(\xi, \boldsymbol{\beta}) \end{bmatrix}} \end{matrix}.$$

Then $\boldsymbol{B}(\boldsymbol{\xi}, \boldsymbol{\beta})$ is the covariance matrix of the $p \times 1$ estimator $\hat{\boldsymbol{\beta}}$, $\boldsymbol{B}_{11}(\xi, \boldsymbol{\beta})$ is the covariance matrix of the $p_1 \times 1$ estimator $\hat{\boldsymbol{\beta}}_1$ in the *presence* of the $p_2 \times 1$ estimator $\hat{\boldsymbol{\beta}}_2$ (i.e., when we fit the full model $\eta = \boldsymbol{f}(\boldsymbol{x})^\top \boldsymbol{\beta}$), and $\boldsymbol{M}_{11}^{-1}(\xi, \boldsymbol{\beta})$ is the covariance matrix of $\hat{\boldsymbol{\beta}}_1$ in the *absence* of $\hat{\boldsymbol{\beta}}_2$ (i.e., when we fit the reduced model $\eta = \boldsymbol{f}_1(\boldsymbol{x})^\top \boldsymbol{\beta}_1$). In the present situation, the D_s-optimal design is the one that minimises $\det[\boldsymbol{B}_{11}(\xi, \boldsymbol{\beta})]$.

Important Note

Except in special circumstances, the matrices $\boldsymbol{B}_{11}(\xi,\boldsymbol{\beta})$ and $\boldsymbol{M}_{11}^{-1}(\xi,\boldsymbol{\beta})$ are not equal. If your experience of matrices is limited, and the previous sentence seems counterintuitive, consider the example of the matrix

$$\boldsymbol{A} = \begin{bmatrix} 1 & 2 \\ 3 & 4 \end{bmatrix},$$

where $p_1 = p_2 = 1$. The matrices

$$\boldsymbol{A}^{-1} = \begin{bmatrix} -2 & 1 \\ 1.5 & -0.5 \end{bmatrix} \quad \text{and} \quad \begin{bmatrix} 1^{-1} & 2^{-1} \\ 3^{-1} & 4^{-1} \end{bmatrix} = \begin{bmatrix} 1 & 0.5 \\ 0.\dot{3} & 0.25 \end{bmatrix}$$

(and their upper left-hand elements in particular) are not equal.

By Harville (1997, Corollary 8.5.12),

$$\boldsymbol{B}_{11}(\xi,\boldsymbol{\beta}) = \left[\boldsymbol{M}_{11}(\xi,\boldsymbol{\beta}) - \boldsymbol{M}_{12}(\xi,\boldsymbol{\beta})\boldsymbol{M}_{22}^{-1}(\xi,\boldsymbol{\beta})\boldsymbol{M}_{21}(\xi,\boldsymbol{\beta})\right]^{-1}.$$

By adapting Harville (1997, Theorem 13.3.8), it follows further that

$$\begin{aligned} \det\left[\boldsymbol{M}(\xi,\boldsymbol{\beta})\right] &= \det\left[\boldsymbol{B}_{11}^{-1}(\xi,\boldsymbol{\beta})\right] \times \det\left[\boldsymbol{M}_{22}(\xi,\boldsymbol{\beta})\right] \\ &= \left\{\det\left[\boldsymbol{B}_{11}(\xi,\boldsymbol{\beta})\right]\right\}^{-1} \times \det\left[\boldsymbol{M}_{22}(\xi,\boldsymbol{\beta})\right], \end{aligned}$$

which implies that

$$\left\{\det\left[\boldsymbol{B}_{11}(\xi,\boldsymbol{\beta})\right]\right\}^{-1} = \det\left[\boldsymbol{M}(\xi,\boldsymbol{\beta})\right]/\det\left[\boldsymbol{M}_{22}(\xi,\boldsymbol{\beta})\right].$$

As $\det[\boldsymbol{B}_{11}(\xi,\boldsymbol{\beta})]$ is minimised by maximising $\det\{[\boldsymbol{B}_{11}(\xi,\boldsymbol{\beta})]\}^{-1}$, then the D_S-optimal design is that one which maximises $\det\left[\boldsymbol{M}(\xi,\boldsymbol{\beta})\right]/\det\left[\boldsymbol{M}_{22}(\xi,\boldsymbol{\beta})\right]$.

Example 3.7.3. *Consider the standard quadratic regression, where the link function is $g(\mu) = \mu$, with $g(\mu) = \eta = \beta_0+\beta_1 x+\beta_2 x^2 = \boldsymbol{f}(\boldsymbol{x})^\top\boldsymbol{\beta}$, where $\boldsymbol{f}(x) = (1, x, x^2)^\top$. We assume that the response variable is normally distributed with var$(Y) = \sigma^2$ for all $i = 1,\ldots,n$. Then the variance function satisfies $V(\mu) = 1$, as shown immediately below* (1.12)*. The model weight was defined in* (3.5)*. As $(\partial\mu)/(\partial\eta) = 1$, the model weight at any support point, $\boldsymbol{x}$, of a design is given by $\omega(x) = 1/\sigma^2$, and it follows from* (3.6) *that the design ξ given by* (3.1) *has the information matrix*

$$M(\xi,\boldsymbol{\beta}) = \sum_{i=1}^{s} \delta_i \frac{1}{\sigma^2}\boldsymbol{f}(x_i)\boldsymbol{f}^\top(x_i) = \frac{1}{\sigma^2}\sum_{i=1}^{s}\delta_i \begin{bmatrix} 1 & x_i & x_i^2 \\ x_i & x_i^2 & x_i^3 \\ x_i^2 & x_i^3 & x_i^4 \end{bmatrix}. \tag{3.19}$$

Let the design space from which support points are selected be $\mathcal{X} = \{x : -1 \leq x \leq 1\}$. As σ^2 is a constant in the definition of $M(\xi,\boldsymbol{\beta})$ above, set it equal to 1. Then it is straightforward, by application of Program_9 from the Web site doeforglm.com, to find that the D-optimal design for the quadratic model seems to be

$$\xi_D^* = \left\{ \begin{array}{ccc} -1 & 0 & 1 \\ \frac{1}{3} & \frac{1}{3} & \frac{1}{3} \end{array} \right\}.$$

This design would be used if it was believed that the appropriate relationship between η and x is quadratic. However, if we were unsure about this, thinking that the relationship might perhaps be a straight line, we might want to place more emphasis on the estimation of β_2. In this case, it would be appropriate to use D_S-optimality. Following (3.18), *write* $\eta = \boldsymbol{f}_1(x)^\top \boldsymbol{\beta}_1 + \boldsymbol{f}_2(x)^\top \boldsymbol{\beta}_2$, *where* $\boldsymbol{\beta}_1 = \beta_2$, $\boldsymbol{f}_1(x) = x^2$, $\boldsymbol{\beta}_2 = (\beta_0, \beta_1)^\top$ *and* $\boldsymbol{f}_2(x) = (1, x)^\top$. *Still using* $\sigma^2 = 1$, *it follows that*

$$\boldsymbol{M}_{22}(\xi, \boldsymbol{\beta}) = \sum_{i=1}^{s} \delta_i \begin{bmatrix} 1 & x_i \\ x_i & x_i^2 \end{bmatrix}. \qquad (3.20)$$

One can use (3.19) *and* (3.20) *to calculate* $det[\boldsymbol{M}(\xi, \boldsymbol{\beta})] / det[\boldsymbol{M}_{22}(\xi, \boldsymbol{\beta})]$. *Then straightforward application of Program_10 suggests that the D_S-optimal design is*

$$\xi^*_{D_s} = \left\{ \begin{array}{ccc} -1 & 0 & 1 \\ 0.25 & 0.5 & 0.25 \end{array} \right\}.$$

By comparison of this design with ξ^*_D above, one sees that D_s-optimality places greater weight on the central support point when the greatest interest is on estimating the quadratic coefficient, β_2.

A further example of finding a D_S-optimal design, for a logistic regression, appears in Sub-section 4.9.

3.7.6 E-optimality

Amongst all designs in a set Ξ, the E-optimal design ξ^*_E is that one that minimises the maximum variance of a linear combination of parameter estimates, $\boldsymbol{a}^\top \hat{\boldsymbol{\beta}}$, where $\boldsymbol{a}^\top \boldsymbol{a} = 1$. That is, for a given design, we ask what is the linear combination of parameters, $\boldsymbol{a}^\top \boldsymbol{\beta}$, that is worst estimated in the sense that its variance is greater than the variance of any other linear combination of parameters, and we choose that particular design for which this maximum variance is least. As it is always possible to find an estimator that has a larger variance by increasing the values of the coefficients (e.g., $\text{var}(2\hat{\beta}_0 + 2\hat{\beta}_1 + 4\hat{\beta}_2)$ has a value that is four times as great as that of $\text{var}(\hat{\beta}_0 + \hat{\beta}_1 + 2\hat{\beta}_2)$), we need some way of ensuring that, when we compare variances, we compare "like with like." This is done by requiring that $\boldsymbol{a}^\top \boldsymbol{a} = \sum_i a_i^2 = 1$; that is, we consider only vectors of coefficients whose sums of squares equal 1. This ensures that we are not increasing a variance simply by increasing the coefficients in $\boldsymbol{a}$.

For a given design ξ, $\text{cov}(\hat{\boldsymbol{\beta}}) = \boldsymbol{M}^{-1}(\xi, \boldsymbol{\beta})$, and so $\text{var}(\boldsymbol{a}^\top \hat{\boldsymbol{\beta}}) = \boldsymbol{a}^\top \boldsymbol{M}^{-1}(\xi, \boldsymbol{\beta}) \boldsymbol{a}$, which is a quadratic form. (See page 27.) We know from the discussion on quadratic forms that, for $\boldsymbol{a}$ satisfying $\boldsymbol{a}^\top \boldsymbol{a} = 1$, the maximum value of the quadratic form is the maximum eigenvalue of $\boldsymbol{M}^{-1}(\xi, \boldsymbol{\beta})$. However, from page 27 the eigenvalues of $\boldsymbol{M}^{-1}(\xi, \boldsymbol{\beta})$ are the reciprocals of the eigenvalues of $\boldsymbol{M}(\xi, \boldsymbol{\beta})$. In all but the most exceptional circumstances, all the eigenvalues of $\boldsymbol{M}(\xi, \boldsymbol{\beta})$ are positive, so the maximum eigenvalue of $\boldsymbol{M}^{-1}(\xi, \boldsymbol{\beta})$ is the reciprocal of the minimum eigenvalue of $\boldsymbol{M}(\xi, \boldsymbol{\beta})$, $\lambda_{\min}^{-1}$. So, for a given design ξ, the maximum variance is $\lambda_{\min}^{-1}$, and E-optimality requires us to select the design that minimises $\lambda_{\text{min}}^{-1}$. However, minimising a reciprocal is equivalent to maximising the original quantity, and so E-optimality requires us to select that design which maximises the minimum eigenvalue of $\boldsymbol{M}(\xi, \boldsymbol{\beta})$.

3.7.7 G-optimality

The standardised variance was first introduced in Sub-section 3.4.2. It is a scaled measure of the variance of the predictor, $\hat{y}$, of the value of y at a specified value of $\boldsymbol{x}$. So far it has been considered only in the context of the general linear model, but it will shortly be generalised for use with GLMs. For a given design ξ, the standardised variance can be calculated for each value of $\boldsymbol{x}$ in the design space. The maximum of all these standardised variances can then be found, and this can be used to characterise the design. The particular design for which this maximum is least is the G-optimal design. This design can be thought of as minimising the width of the widest confidence interval for the value of y.

3.8 Example

Suppose that we want to find a D-optimal design for the basic logistic regression with two model parameters and one explanatory variable:

$$\text{logit}(\pi) = \ln\left(\frac{\pi}{1-\pi}\right) = \beta_0 + \beta_1 x.$$

As there are two parameters, the minimum number of support points that might be required is $p = 2$, and the maximum number is $p(p+1)/2 = 3$. The D-optimality criterion requires the maximisation of $\det[\boldsymbol{M}(\xi, \boldsymbol{\beta})]$, which depends on the vector of parameters, $\boldsymbol{\beta}$, that is used in the calculation of $\boldsymbol{M}(\xi, \boldsymbol{\beta})$. The resulting D-optimal design is said to be *locally optimal,* as it is applicable only to this particular value of $\boldsymbol{\beta}$.

This threatens to require a separate search for a design for each value of $\boldsymbol{\beta}$ in which we are interested. To get around this difficulty for this particular model, we use the transformation

$$z = \beta_0 + \beta_1 x. \tag{3.21}$$

This was introduced by Ford, Torsney & Wu (1992), who called it the *canonical transformation.* Its advantage is that now $\text{logit}(\pi) = z$, which does not depend on any model parameters. Then it follows that the model weight is given by

$$\begin{aligned}
\omega(z) &= \pi(z)[1-\pi(z)] \qquad (3.22)\\
&= \frac{1}{\exp(-z)+1} \times \frac{\exp(-z)}{\exp(-z)+1}\\
&= \frac{\exp(-z)}{[\exp(-z)+1]^2} \times \frac{\exp(2z)}{\exp(2z)} \text{ (for simplification purposes)}\\
&= \frac{\exp(z)}{[1+\exp(z)]^2}. \qquad (3.23)
\end{aligned}$$

The canonical transformation lets us write $\boldsymbol{f}(\boldsymbol{z}) = \boldsymbol{B}\boldsymbol{f}(\boldsymbol{x})$, where

$$\boldsymbol{f}(\boldsymbol{z}) = \begin{bmatrix} 1 \\ z \end{bmatrix}, \quad \boldsymbol{B} = \begin{bmatrix} 1 & 0 \\ \beta_0 & \beta_1 \end{bmatrix} \text{ and } \boldsymbol{f}(\boldsymbol{x}) = \begin{bmatrix} 1 \\ x \end{bmatrix}.$$

By Result 2.2.2, the determinant of $\boldsymbol{B}$ is $\det(\boldsymbol{B}) = 1 \times \beta_1 - \beta_0 \times 0 = \beta_1$. For $\beta_1 \neq 0$, $\det(\boldsymbol{B}) \neq 0$, and so $\boldsymbol{B}^{-1}$ exists. Hence

$$\boldsymbol{f}(\boldsymbol{x}) = \boldsymbol{B}^{-1}\boldsymbol{f}(\boldsymbol{z}). \tag{3.24}$$

Consider a design formulated in terms of the variable x:

$$\xi_x = \left\{ \begin{array}{cccc} x_1 & x_2 & \dots & x_s \\ \delta_1 & \delta_2 & \dots & \delta_s \end{array} \right\}.$$

Suppose that the canonical transformation maps x_i to z_i $(i = 1, \dots, s)$. The transformation has no effect on the design weight at each point. So, in terms of the transformed variable z, the design ξ_x may now be written as

$$\xi_z = \left\{ \begin{array}{cccc} z_1 & z_2 & \dots & z_s \\ \delta_1 & \delta_2 & \dots & \delta_s \end{array} \right\}.$$

The information matrix for ξ_x is given by

$$\begin{aligned} \boldsymbol{M}(\xi_x, \boldsymbol{\beta}) &= \sum_{i=1}^{s} \delta_i\, \omega(x_i) \boldsymbol{f}(x_i) \boldsymbol{f}^{\top}(x_i) \\ &= \sum_{i=1}^{s} \delta_i \frac{\exp(\beta_0 + \beta_1 x_i)}{[1 + \exp(\beta_0 + \beta_1 x_i)]^2} \left[\boldsymbol{B}^{-1}\boldsymbol{f}(z_i)\right] \left[\boldsymbol{B}^{-1}\boldsymbol{f}(z_i)\right]^{\top} \\ &= \sum_{i=1}^{s} \delta_i\, \omega(z_i) \boldsymbol{B}^{-1}\boldsymbol{f}(z_i) \left[\boldsymbol{B}^{-1}\boldsymbol{f}(z_i)\right]^{\top} \\ &= \boldsymbol{B}^{-1} \sum_{i=1}^{s} \delta_i\, \omega(z_i) \boldsymbol{f}(z_i) \boldsymbol{f}^{\top}(z_i) \left(\boldsymbol{B}^{-1}\right)^{\top} \\ &= \boldsymbol{B}^{-1} \boldsymbol{M}(\xi_z) \left(\boldsymbol{B}^{-1}\right)^{\top}, \end{aligned}$$

using (3.22) and (3.24), and where $\boldsymbol{M}(\xi_z)$ is the information matrix corresponding to the transformed design ξ_z.

Taking determinants of both sides of this equation and using results on page 26 gives

$$\begin{aligned} \det\left[\boldsymbol{M}(\xi_x, \boldsymbol{\beta})\right] &= \det\left[\boldsymbol{B}^{-1} \boldsymbol{M}(\xi_z) \left(\boldsymbol{B}^{-1}\right)^{\top}\right] \\ &= \det(\boldsymbol{B}^{-1}) \times \det\left[\boldsymbol{M}(\xi_z)\right] \times \det\left[\left(\boldsymbol{B}^{-1}\right)^{\top}\right] \\ &= \left[\det(\boldsymbol{B})\right]^{-1} \times \det\left[\boldsymbol{M}(\xi_z)\right] \times \left[\det(\boldsymbol{B})\right]^{-1} \\ &= (1/\beta_1)^2 \det\left[\boldsymbol{M}(\xi_z)\right]. \end{aligned}$$

So $\det[\boldsymbol{M}(\xi_x,\boldsymbol{\beta})]$ is just a constant multiple of $\det[\boldsymbol{M}(\xi_z)]$, and it follows that maximising $\det[\boldsymbol{M}(x,\boldsymbol{\beta})]$ over a specified set of x-values is equivalent to maximising $\det[\boldsymbol{M}(\xi_z)]$ over the set of z-values that correspond to the transformed x-values. That is, we can find the locally D-optimal design for $\boldsymbol{\beta} = (\beta_0, \beta_1)^\top$ by finding the globally D-optimal design for z.

3.8.1 Using constrOptim

The function below, *infodet*, has as input a vector $(z_1, \ldots, z_s, \delta_1, \ldots, \delta_s)$ for an arbitrary value of s. It then calculates $-\det[\boldsymbol{M}(\xi)]$ (as maximising $\det[\boldsymbol{M}(\xi)]$ is the same as minimising $-\det[\boldsymbol{M}(\xi)]$). The function can use any values of s, β_0 and β_1, which are specified outside the program. To use this program for z, set $\beta_0 = 0$ and $\beta_1 = 1$.

```
infodet <- function(x)
{
 info <- matrix(0,2,2)
 for (i in 1:s)
{
  pt <- x[i]
  delta <- x[i+s]
  expeta <- exp(beta0 + beta1*pt)
  wt <- expeta/(1+expeta)^2
  info <- info + delta*wt*matrix(c(1,pt,pt,pt^2),2,2)
 }
 -det(info)
}
```

This version of *infodet* is specific to this particular problem, and will run rather slowly because of the presence of the loop. A faster and more general version of *infodet* will be provided soon.

If we have absolutely no idea of an appropriate domain for the value of z, we might choose something like $-10 \leq z_i \leq 10$ $(i = 1, \ldots, s)$. Additionally, the design weights must satisfy $0 < \delta_i < 1$ and $\delta_1 + \cdots + \delta_s = 1$. Following Example 2.4.1, these constraints may be written in the matrix form $\boldsymbol{C}\boldsymbol{v} - \boldsymbol{u} \geq \boldsymbol{0}$, where $\boldsymbol{v} = (z_1, \ldots, z_s, \delta_1, \ldots, \delta_s)^\top$, by choosing

$$\boldsymbol{C} = \begin{bmatrix} \boldsymbol{I}_s & \boldsymbol{0}_{s\times s} \\ -\boldsymbol{I}_s & \boldsymbol{0}_{s\times s} \\ \boldsymbol{0}_{s\times s} & \boldsymbol{I}_s \\ \boldsymbol{0}_{s\times s} & -\boldsymbol{I}_s \\ \boldsymbol{0}_s^\top & -\boldsymbol{1}_s^\top \end{bmatrix} \quad \text{and} \quad \boldsymbol{u} = \begin{bmatrix} -10\,\boldsymbol{1}_s \\ -10\,\boldsymbol{1}_s \\ \boldsymbol{0}_s \\ -\boldsymbol{1}_s \\ 1 \end{bmatrix}.$$

There are more block rows in this matrix $\boldsymbol{C}$ than in the matrix on page 34 because there are more types of constraints in the current problem.

As a D-optimal design is sought for $p = 2$ parameters, I commenced with the upper bound of $p(p+1)/2 = 3$ support points. I chose $(-1, 0, 1, 0.33, 0.33, 0.33)$ as my initial guess of the solution vector. (Remember that the initial guesses must not lie on the boundary of the solution set, so I must not choose values of $\delta_1, \ldots, \delta_s$ that add to 1.) I used the following program:

```
s <- 3
beta0 <- 0
beta1 <- 1
i3 <- diag(1,3)
cmat1 <- rbind(cbind(i3,0*i3),cbind(-i3,0*i3))
cmat2 <- rbind(cbind(0*i3,i3),cbind(0*i3,-i3))
cmat <- rbind(cmat1,cmat2,c(rep(0,3),rep(-1,3)))
uvec <- c(rep(-10,6),rep(0,3),rep(-1,3),-1)
start <- c(-1,0,1,0.33,0.33,0.33)
out <- constrOptim(start,infodet,NULL,cmat,uvec,
   method="Nelder-Mead")
out
```

Note that I did not attempt to find the derivatives of $-\det[\boldsymbol{M}(\xi)]$ with respect to each of $z_1, \ldots, z_s$ and $\delta_1, \ldots, \delta_s$, but instead used the Nelder-Mead algorithm. (See page 34.)

The output gave the following for the values of $(z_1, z_2, z_3, \delta_1, \delta_2, \delta_3)$ for the D-optimal design:

```
-1.545534e+00  3.236333e-01  1.541277e+00  4.859467e-01
 1.423844e-08  5.140533e-01
```

The minimised value of $-\det[\boldsymbol{M}(\xi)]$ was -0.05007884.

The value of δ_2 is 1.42×10^{-8}, which is essentially zero, suggesting that only two support points are needed. Note that z_1 is almost equal to $-z_3$ and δ_1 and δ_3 are both approximately 0.5. This suggests that an "elegant" solution might be forthcoming if we try just two support points.

The following commands were introduced to replace the last nine commands above:

```
s <- 2
i2 <- diag(1,2)
cmat1 <- rbind(cbind(i2,0*i2),cbind(-i2,0*i2))
cmat2 <- rbind(cbind(0*i2,i2),cbind(0*i2,-i2))
cmat <- rbind(cmat1,cmat2,c(rep(0,2),rep(-1,2)))
uvec <- c(rep(-10,4),rep(0,2),rep(-1,2),-1)
start <- c(-1.5,1.5,0.495,0.495)
out <- constrOptim(start,infodet,NULL,cmat,uvec,
   method="Nelder-Mead")
out
```

This gave $z_1 = -1.5435353$, $z_2 = 1.5435583$, $\delta_1 = 0.5000248$, $\delta_2 = 0.4999752$, and the minimised value of $-\det[\boldsymbol{M}(\xi)]$ was equal to -0.05011849. The value of $\det[\boldsymbol{M}(\xi)]$ has slightly increased, from 0.05007884 to 0.05011849, as a result of decreasing the number of support points. Allowing for numerical noise, the overall result suggests that the D-optimal design is

$$\xi_z = \left\{ \begin{array}{cc} -1.5435 & 1.5435 \\ 0.5 & 0.5 \end{array} \right\}.$$

This result can be examined mathematically by assuming that the optimal design is of the form

$$\xi_z = \left\{ \begin{array}{cc} -a & a \\ 0.5 & 0.5 \end{array} \right\}.$$

for some value of a still to be determined. Then

$$\begin{aligned} \boldsymbol{M}(\xi_z) &= \sum_{i=1}^{2} \delta_i \frac{\exp(z_i)}{[1+\exp(z_i)]^2} \left[\begin{array}{cc} 1 & z_i \\ z_i & z_i^2 \end{array} \right] \\ &= \frac{1}{2} \left\{ \frac{\exp(-a)}{[1+\exp(-a)]^2} \left[\begin{array}{cc} 1 & -a \\ -a & (-a)^2 \end{array} \right] + \frac{\exp(a)}{[1+\exp(a)]^2} \left[\begin{array}{cc} 1 & a \\ a & a^2 \end{array} \right] \right\} \\ &= \frac{\exp(a)}{[1+\exp(a)]^2} \left[\begin{array}{cc} 1 & 0 \\ 0 & a^2 \end{array} \right]. \end{aligned} \tag{3.25}$$

This uses the result that $\exp(-a)/[1+\exp(-a)]^2 = \exp(a)/[1+\exp(a)]^2$, as illustrated in (3.22).

It now follows directly that

$$\det\left[\boldsymbol{M}(\xi_z)\right] = \left\{ \frac{\exp(a)}{[1+\exp(a)]^2} \right\}^2 \times a^2 = \frac{a^2 \exp(2a)}{[1+\exp(a)]^4}.$$

This determinant is clearly a function, $h(a)$ (say), of a. This function

- satisfies $h(a) = h(-a)$ (i.e., it is an even function), so it is symmetric around $a = 0$;
- satisfies $\lim_{a \to \infty} h(a) = 0$.

Consequently, when exploring the behaviour of this function, we can be confident that there will not be maxima occurring at large values of a. A graph of $h(a)$ vs a appears in Figure 3.6, and it strongly suggests that $h(a)$ has only two equal-valued maxima, at approximately $a = \pm 1.55$. The following program defines the function $h(a)$, and then uses the R function `optimise` (which can also be spelled `optimize`) to find where the maximum of $h(a)$ occurs for $a \in \{a : 0 \leq a \leq 10\}$.

```
h <- function(a)
{
  expa <- exp(a)
  f <- (a^2)*(expa^2)/((1+expa)^4)
  f
}
out <- optimise(h,c(0,10),maximum=TRUE)
out
```

It gave the output

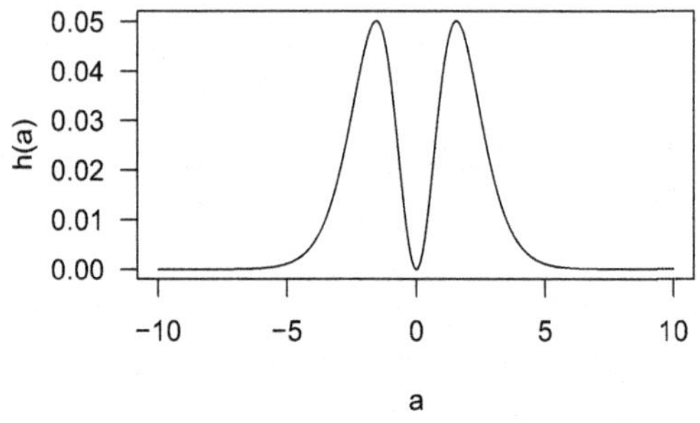

Figure 3.6 *Graph of* $h(a) = a^2 \exp(2a)/[1 + \exp(a)]^4$ *vs* a *for* $-10 \leq a \leq 10$.

```
$maximum
[1] 1.543385
$objective
[1] 0.05011849
```

So we conclude that the D-optimal design for the canonical variable z is

$$\xi_z = \left\{ \begin{array}{cc} -1.5434 & 1.5434 \\ 0.5 & 0.5 \end{array} \right\}. \tag{3.26}$$

3.8.2 Using optim

We can use *optim* rather than *constrOptim* by making some small changes to the function *infodet* on page 77. These will allow values of z as input, and by appropriate choice of transformations, the z-values can produce values $x_1, \ldots, x_s$ and $\delta_1, \ldots, \delta_s$ that satisfy the constraints $-10 \leq x_i \leq 10$ and $\delta_i > 0$; $\delta_1 + \cdots + \delta_s = 1$. To obtain the x-values, I used the fact that $0 < z < 1 \Rightarrow -10 < 10\cos(\pi z) < 10$ (a minor modification of Method 3 on page 38). The vector of weights has been generated directly using Method 5 on page 38.

For the initial z-values, I chose to input $2s$ values from the distribution that is uniform on $(0, 1)$. The program is

```
infodet <- function(z)
{
 xvals <- 10*cos(pi*z[1:s])
 temp <- (z[(s+1):(2*s)])^2
 deltavec <- temp/(sum(temp))
 info <- matrix(0,2,2)
 for (i in 1:s)
{
  pt <- xvals[i]
  delta <- deltavec[i]
```

```
  expeta <- exp(beta0 + beta1*pt)
  wt <- expeta/(1+expeta)^2
  info <- info + delta*wt*matrix(c(1,pt,pt,pt^2),2,2)
 }
 -det(info)
}

beta0 <- 0
beta1 <- 1
s <- 3

start <- runif(2*s)
out <- optim(start,infodet,NULL,method="Nelder-Mead")
z <- out$par
xvals <- 10*cos(pi*z[1:s])
temp <- (z[(s+1):(2*s)])^2
deltavec <- temp/(sum(temp))
rbind(xvals,deltavec)
```

Remember that the input to, and output from, *optim* are values of z. Once an optimal design has been found, the output must be converted into values of x_i and δ_i $(i = 1, \ldots, s)$.

This program runs more quickly than *constrOptim*. The output that I obtained from a run of the program was

```
              [,1]     [,2]         [,3]
xvals    -1.543403 1.543407 8.691659e+00
deltavec  0.500000 0.500000 1.420785e-13
```

As we saw with *constrOptim*, the very small value of δ_3 suggests that only two support points are needed. One can simply replace `s <- 3` by `s <- 2` and run subsequent lines of the program again. (This is much easier than having to change the specification of the constraints when using *constrOptim*.)

Running the above program on a different occasion gave the following output:

```
               [,1]       [,2]      [,3]
xvals    -1.5434031 -1.5433998 1.5434069
deltavec  0.2259099  0.2740888 0.5000013
```

Although this output appears to suggest the need for three support points, note that z_1 and z_2 are almost identical, and that $\delta_1 + \delta_2 \approx 0.5$. So in fact this leads to the same conclusion as the previous output from this program: that only $s = 2$ support points are required.

Running the program for $s = 2$ gave the output

```
               [,1]      [,2]
xvals    -1.5434044 1.5434047
deltavec  0.4999998 0.5000002
```

which agrees with the results obtained earlier.

3.8.3 *How to be sure that you have the right design*

By making reasonable assumptions from the results of *constrOptim* or *optim*, we have arrived at what we believe to be the D-optimal design for a logit link with $\eta = z$. However, what if these assumptions are not correct? Or what if we have a much more complicated problem, with several predictor variables, and *constrOptim* or *optim* gives a solution that does not have any evident pattern amongst the design weights or support points? How can we then be sure that a design we obtain is D-optimal? The answer to this question is the topic of the next section.

3.9 The general equivalence theorem

The mathematics that underlies the results in this section is very much beyond the scope of this book. References are provided for those who wish to follow the mathematics. For those who are happy to accept that someone else has proven the results, the explanation should be sufficient for you to see what the results say, why they are important, and how they can be applied.

We saw in the sub-sections on A-, D-, D$_s$- and E-optimality that the optimality criteria involved minimising some function $\psi[\boldsymbol{M}(\xi, \boldsymbol{\beta})]$ of the information matrix of a design, ξ. For example, A-optimality required the minimisation of $\mathrm{tr}\left[\boldsymbol{M}^{-1}(\xi, \boldsymbol{\beta})\right]$, while D-optimality required the minimisation of $\det\left[\boldsymbol{M}^{-1}(\xi, \boldsymbol{\beta})\right]$. Although the aim is to find the minimum of $\psi\left[\boldsymbol{M}(\xi, \boldsymbol{\beta})\right]$, sometimes ψ is not of an appropriate form, and it is necessary to minimise a function of ψ. For example, for D-optimality, while the criterion says to minimise $\det\left[\boldsymbol{M}^{-1}(\xi, \boldsymbol{\beta})\right]$, it is better to minimise

$$\ln\left\{\det\left[\boldsymbol{M}^{-1}(\xi, \boldsymbol{\beta})\right]\right\} = \ln\left(\{\det\left[\boldsymbol{M}(\xi, \boldsymbol{\beta})\right]\}^{-1}\right) = -\ln\left\{\det\left[\boldsymbol{M}(\xi, \boldsymbol{\beta})\right]\right\}$$

instead.

While you will have seen in introductory calculus the notion of a derivative of a function of one or more variables, you may not have seen the concept of the derivative of a function of a matrix. Consider the design ξ in (3.1), and now consider an alternative design

$$\bar{\xi}_1 = \left\{ \begin{array}{c} \boldsymbol{x} \\ 1 \end{array} \right\}$$

that has just the one support point at $\boldsymbol{x}$. For a given value of $\boldsymbol{\beta}$, these two designs have information matrices $\boldsymbol{M}(\xi, \boldsymbol{\beta})$ and $\boldsymbol{M}(\xi_1, \boldsymbol{\beta})$, respectively. If we now form a new design

$$\xi_\alpha = (1 - \alpha)\xi + \alpha\xi_1$$

that gives a weighting of $(1-\alpha)$ to ξ and a weighting of α to ξ_1 (for $0 \leq \alpha \leq 1$), then

$$\xi_\alpha = \left\{ \begin{array}{cccccc} \boldsymbol{x}_1 & \boldsymbol{x}_2 & \dots & \boldsymbol{x}_s & \boldsymbol{x} \\ (1-\alpha)\delta_1 & (1-\alpha)\delta_2 & \dots & (1-\alpha)\delta_s & \alpha \end{array} \right\},$$

and the information matrix of ξ_α is

$$\begin{aligned}
\boldsymbol{M}(\xi_\alpha, \boldsymbol{\beta}) &= \sum_{i=1}^{s}(1-\alpha)\delta_i\, \omega(\boldsymbol{x}_i)\boldsymbol{f}(\boldsymbol{x}_i)\boldsymbol{f}^\top(\boldsymbol{x}_i) + \alpha\omega(\boldsymbol{x})\boldsymbol{f}(\boldsymbol{x})\boldsymbol{f}^\top(\boldsymbol{x}) \\
&= (1-\alpha)\times\sum_{i=1}^{s} \delta_i\, \omega(\boldsymbol{x}_i)\boldsymbol{f}(\boldsymbol{x}_i)\boldsymbol{f}^\top(\boldsymbol{x}_i) + \alpha\times 1\, \omega(\boldsymbol{x})\boldsymbol{f}(\boldsymbol{x})\boldsymbol{f}^\top(\boldsymbol{x}) \\
&= (1-\alpha)\boldsymbol{M}(\xi, \boldsymbol{\beta}) + \alpha\boldsymbol{M}(\xi_1, \boldsymbol{\beta}). \qquad (3.27)
\end{aligned}$$

We want to see how the value of $\psi\left[\boldsymbol{M}(\xi, \boldsymbol{\beta})\right]$ alters as we move from ξ in the direction of the design ξ_1. Essentially, we want the derivative of $\psi\left[\boldsymbol{M}(\xi, \boldsymbol{\beta})\right]$ in the direction of ξ_1. The change in values of $\psi(\cdot)$ is

$$\begin{aligned}
&\psi\left[\boldsymbol{M}(\xi_\alpha, \boldsymbol{\beta})\right] - \psi\left[\boldsymbol{M}(\xi, \boldsymbol{\beta})\right] \\
=&\psi\left[(1-\alpha)\boldsymbol{M}(\xi, \boldsymbol{\beta}) + \alpha\boldsymbol{M}(\xi_1, \boldsymbol{\beta})\right] - \psi\left[\boldsymbol{M}(\xi, \boldsymbol{\beta})\right].
\end{aligned}$$

Dividing this difference by α will give the gradient of the change resulting from including a proportion α of ξ_1 in the design ξ_α, and taking the limit of this ratio as α approaches 0 from the positive side (denoted by $\alpha \to 0^+$) will give a derivative: the rate of change in $\psi\left[\boldsymbol{M}(\xi, \boldsymbol{\beta})\right]$ in the direction of ξ_1. Denote this derivative by $\phi(\boldsymbol{x}, \xi, \boldsymbol{\beta})$. Then

$$\phi(\boldsymbol{x}, \xi, \boldsymbol{\beta}) = \lim_{\alpha \to 0^+} \frac{1}{\alpha}\left\{\psi\left[(1-\alpha)\boldsymbol{M}(\xi, \boldsymbol{\beta}) + \alpha\boldsymbol{M}(\xi_1, \boldsymbol{\beta})\right] - \psi\left[\boldsymbol{M}(\xi_\alpha, \boldsymbol{\beta})\right]\right\}.$$

The quantity $\phi(\boldsymbol{x}, \xi, \boldsymbol{\beta})$ is known as the *Fréchet derivative.* Clearly it depends on the design ξ and the parameter vector $\boldsymbol{\beta}$. However, for the investigation of a candidate design, ξ, the values of ξ and $\boldsymbol{\beta}$ are fixed, and the derivative is regarded as a function of $\boldsymbol{x}$, the support point of the design ξ_1.

Let us consider how we would expect the derivative $\phi(\boldsymbol{x}, \xi, \boldsymbol{\beta})$ to behave at different values of $\boldsymbol{x}$.

1. If $\boldsymbol{x}$ is equal to one of the support points of ξ, we would expect no change in $\psi\left[\boldsymbol{M}(\xi, \boldsymbol{\beta})\right]$; that is, the gradient would be zero. This is true whether or not ξ is the optimal design.
2. If ξ is the optimal design, i.e., it minimises $\psi\left[\boldsymbol{M}(\xi, \boldsymbol{\beta})\right]$, then a movement in the direction of any non-support point would in fact increase $\psi\left[\boldsymbol{M}(\xi, \boldsymbol{\beta})\right]$; i.e., the gradient would be positive.
3. If ξ is not the optimal design, so it does not minimise $\psi\left[\boldsymbol{M}(\xi, \boldsymbol{\beta})\right]$, then there will be some potential support points $\boldsymbol{x}$ whose inclusion in ξ would decrease $\psi\left[\boldsymbol{M}(\xi, \boldsymbol{\beta})\right]$. Moving in the direction of ξ_1 would give a negative gradient.

This is the rationale behind the general equivalence theorem.

General equivalence theorem (Kiefer & Wolfowitz, 1960)

The following three conditions are equivalent on the optimal design ξ^*.

1. The design ξ^* minimises $\psi(\xi)$.

2. The minimum value over $\mathcal{X}$ of $\phi(\boldsymbol{x}, \xi, \boldsymbol{\beta})$ is maximised by the design ξ^*.
3. The minimum value over $\mathcal{X}$ of $\phi(\boldsymbol{x}, \xi^*, \boldsymbol{\beta})$ is 0, and this value occurs at each of the support points of ξ^*.

The nature of the Fréchet derivative $\phi(\boldsymbol{x}, \xi, \boldsymbol{\beta})$ depends crucially on the function $\psi(\cdot)$ of $\boldsymbol{M}(\xi, \boldsymbol{\beta})$ that we are trying to minimise.

- For **D-optimality**, where we seek to minimise det $\left[\boldsymbol{M}^{-1}(\xi, \boldsymbol{\beta})\right]$, or equivalently ln $\left\{\det\left[\boldsymbol{M}^{-1}(\xi, \boldsymbol{\beta})\right]\right\}$, the derivative satisfies

$$\phi(\boldsymbol{x}, \xi, \boldsymbol{\beta}) = p - d(\boldsymbol{x}, \xi, \boldsymbol{\beta}),$$

  where

$$d(\boldsymbol{x}, \xi, \boldsymbol{\beta}) = \omega(\boldsymbol{x})\boldsymbol{f}^{\top}(\boldsymbol{x})\boldsymbol{M}^{-1}(\xi, \boldsymbol{\beta})\boldsymbol{f}(\boldsymbol{x})$$

  is the standardised variance of the design ξ and p is the number of parameters in $\boldsymbol{\beta}$.

  It follows from point 3 of the Generalised Equivalence Theorem that $\phi(\boldsymbol{x}, \xi^*, \boldsymbol{\beta}) = p - d(\boldsymbol{x}, \xi^*, \boldsymbol{\beta}) \geq 0$. So $d(\boldsymbol{x}, \xi^*, \boldsymbol{\beta}) \leq p$ everywhere on the design space $\mathcal{X}$, and $d(\boldsymbol{x}, \xi^*, \boldsymbol{\beta}) = p$ at each of the support points. We use this result to test whether a design ξ^* is D-optimal. *If $d(\boldsymbol{x}, \xi^*, \boldsymbol{\beta}) = p$ at each support point of ξ^* (and possibly at other points $\boldsymbol{x} \in \mathcal{X}$) and nowhere on $\mathcal{X}$ is $d(\boldsymbol{x}, \xi^*, \boldsymbol{\beta}) > p$, then ξ^* is D-optimal.*

- It can be shown that the minimum value of the maximum standardised variance for any approximate design is p. We know that the standardised variance of the D-optimal design does not exceed p, so the D-optimal design has minimised the maximum value of the standardised variance. Thus an approximate design which is D-optimal will also be G-optimal.

- For **D$_S$-optimality**, where we seek to minimise det$[\boldsymbol{B}_{11}(\xi, \boldsymbol{\beta})]$,

$$\phi(\boldsymbol{x}, \xi, \boldsymbol{\beta}) = p_1 - d(\boldsymbol{x}, \xi, \boldsymbol{\beta}),$$

  where

$$d(\boldsymbol{x}, \xi, \boldsymbol{\beta}) = \omega(\boldsymbol{x})\left[\boldsymbol{f}^{\top}(\boldsymbol{x})\boldsymbol{M}^{-1}(\xi, \boldsymbol{\beta})\boldsymbol{f}(\boldsymbol{x}) - \boldsymbol{f}_2^{\top}(\boldsymbol{x})\boldsymbol{M}_{22}^{-1}(\xi, \boldsymbol{\beta})\boldsymbol{f}_2(\boldsymbol{x})\right]$$

  is the standardised variance of the design ξ and p_1 is the number of parameters in the subset of $\boldsymbol{\beta}$ that is of interest. (The notation for D$_S$-optimality is that used in Subsection 3.7.5.)

  By the general equivalence theorem, *if $d(\boldsymbol{x}, \xi^*, \boldsymbol{\beta}) = p_1$ at each support point of ξ^* (and possibly at other points $\boldsymbol{x} \in \mathcal{X}$) and nowhere on $\mathcal{X}$ is $d(\boldsymbol{x}, \xi^*, \boldsymbol{\beta}) > p_1$, then ξ^* is D_S-optimal.*

Example 3.9.1. *On page 75, it was conjectured that*

$$\xi_z = \left\{ \begin{array}{cc} -1.5434 & 1.5434 \\ 0.5 & 0.5 \end{array} \right\}$$

is the globally D-optimal design for the canonical variable $z = \beta_0 + \beta_1 x$ for a *logit link and a Bernoulli distribution. It follows from (3.25) that the information matrix for this design is*

$$M(\xi_z) = \frac{\exp(1.5434)}{[1+\exp(1.5434)]^2}\begin{bmatrix} 1 & 0 \\ 0 & 1.5434^2 \end{bmatrix}.$$

The standardised variance at an arbitrary point z is

$$d(z,\xi_z) = \omega(z)\boldsymbol{f}^\top(z)\boldsymbol{M}^{-1}(\xi_z)\boldsymbol{f}(z) = \frac{\exp(z)}{[1+\exp(z)]^2}[1,z]\boldsymbol{M}^{-1}(\xi_z)[1,z]^\top.$$

Note that the standardised variance $d(z,\xi_z)$ does not have an argument $\boldsymbol{\beta}$ because we are dealing with the canonical variable.

The standardised variance is evaluated by the following R commands (that are also available as Program_11 in doeforglm.com):

①

```
beta0 <- 0
beta1 <- 1
s <- 2
```

②

```
infomat <- function(x)
{
 info <- matrix(0,2,2)
 for (i in 1:s)
{
  pt <- x[i]
  delta <- x[i+s]
  expeta <- exp(beta0 + beta1*pt)
  wt <- expeta/(1+expeta)^2
  info <- info + delta*wt*matrix(c(1,pt,pt,pt^2),2,2)
 }
 info
}
```

③

```
optdesign <- c(-1.5434,1.5434,0.5,0.5)
optmat <- infomat(optdesign)
invmat <- solve(optmat)
```

④

```
stdvar <- function(x)
{
 expeta <- exp(x)
 wt <- expeta/(1+expeta)^2
 fx <- matrix(c(1,x),2,1)
 sv <- wt*t(fx)%*%invmat%*%fx
```

```
 sv
}
```

If the design is to be D-optimal, then the standardised variance must equal $p = 2$ at the two support points of the design, and must not exceed two anywhere in the design space $\mathcal{Z} = \{z : -10 < z < 10\}$. The next lines of the program evaluate the standardised variance at the two support points, and plot the standardised variance for all $z \in \mathcal{Z}$. The final four commands draw the dotted lines, and mark the support points.

```
stdvar(-1.5434)
stdvar(1.5434)

x <- seq(from=-10,to=10,by=0.02)
lx <- length(x)
y <- rep(0,lx)
for (i in 1:lx)
{
 y[i] <- stdvar(x[i])
}
par(las=1)
plot(x,y,ty="l",xlab="z",ylab="Standardised Variance",lwd=2,xaxt="n")
axis(1,at=c(-10,-5,-1.5434,1.5434,5,10),
   label=c("-10","-5","-1.5434","1.5434","5","10"))
lines(c(-10,10),c(2,2),lty=2,lwd=2)
lines(c(-1.5434,-1.5434),c(-0.2,2),lty=2,lwd=2)
lines(c(1.5434,1.5434),c(-0.2,2),lty=2,lwd=2)
points(c(-1.5434,1.5434),c(2,2),pch=16,cex=2)
```

The standardised variance is given as 2 for both support points. A plot of the standardised variance for $z \in \mathcal{Z}$ appears in Figure 3.7, and shows that the standardised variance does not exceed p anywhere in the design space. Therefore, by the general equivalence theorem, the design

$$\xi_z = \left\{ \begin{array}{cc} -1.5434 & 1.5434 \\ 0.5 & 0.5 \end{array} \right\}$$

is indeed D-optimal.

If you wish to avoid the loop in the plotting of the standardised variance, the following four commands can replace the commands beginning with `lx <- length(x)` *and ending with* `lines(c(-10,10),c(2,2),lty=2)`*:*

```
y <- sapply(x,stdvar)
par(las=1)
plot(x,y,ty="l",ylab="standardised variance")
lines(c(-10,10),c(2,2),lty=2)
```

Example 3.9.2. *On page 74, it was conjectured that*

$$\xi^*_{D_S} = \left\{ \begin{array}{ccc} -1 & 0 & 1 \\ 0.25 & 0.5 & 0.25 \end{array} \right\}$$

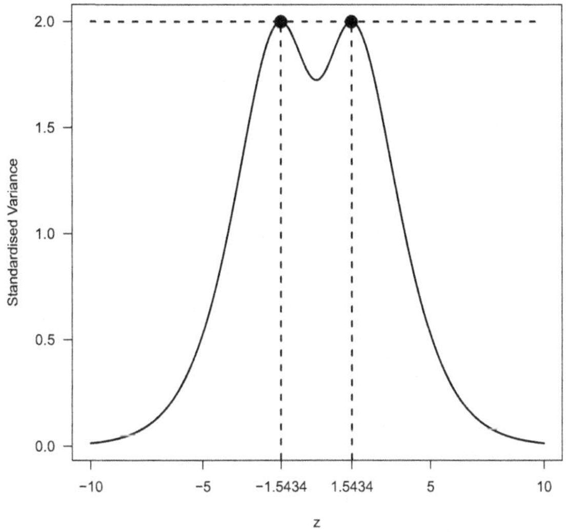

Figure 3.7 *Plot of the standardised variance for the design in Example 3.9.1 believed to be D-optimal on the design space* $\mathcal{Z} = \{z\colon -10 < z < 10\}$.

is the D_S*-optimal design when we are especially interested in the quadratic term in the general linear model*

$$Y_i = \beta_0 + \beta_1 x_i + \beta_2 x_i^2 + E_i.$$

Program_12 calculates the standardised variance, evaluates it at each of the three support points, and plots the standardised variance for all $x \in \{x : -1 < x < 1\}$*. The standardised variance was found to equal* $p_1 = 1$ *at each of the support points (-1, 0 and 1). A plot of the standardised variance vs.* x *appears in Figure 3.8. It can be seen that the standardised variance does not exceed* $p_1 = 1$ *anywhere in the design space. Thus use of the general equivalence theorem confirms that the design* $\xi^*_{D_S}$ *is indeed the* D_S*-optimal design for estimating* β_2 *in the presence of* β_0 *and* β_1 *under the circumstances described above.*

3.10 Where next?

We have now covered the general theory of D- and D_S-optimal designs for GLMs. In the next three chapters, we will look at data from specific distributions, in particular the Bernoulli and Poisson distributions. These are the two distributions most often assumed in work on GLMs, and specific attention will be paid to each distribution in turn.

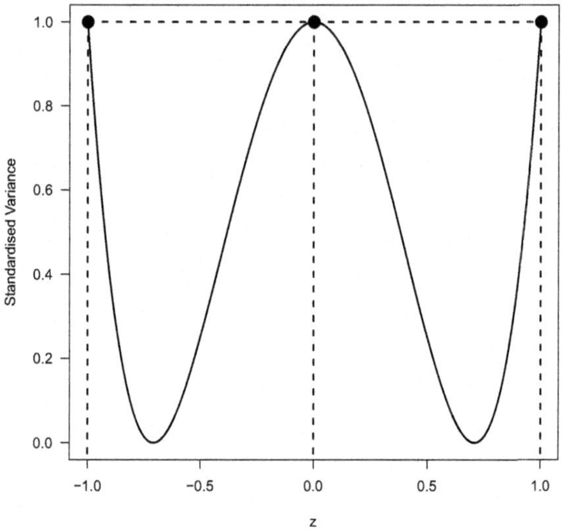

Figure 3.8 *Plot of the standardised variance for the design in Example 3.9.2 believed to be* D_S*-optimal on the design space* $\mathcal{X} = \{x: -1 \leq x \leq 1\}$.

The dependence of designs on particular values of a parameter vector $\boldsymbol{\beta}$ is a distinct disadvantage in designing an experiment. Instead, one might wish to specify a range of values for some or all of the parameters, and maybe to specify a statistical distribution for a parameter. For example, rather than use an estimate of 1 for β_0, one might want to say that it lies between 0.5 and 1.5. The ability to design experiments under such broader assumptions about the values of parameters is a major benefit of a fast-developing area of Experimental Design known as *Bayesian experimental design,* which will be considered in Chapter 7.

Chapter 4

The Binomial Distribution

4.1 Introduction

This chapter contains material on the design of experiments when each observation is assumed to come from the Bernoulli or binomial distributions. The three commonly used link functions, the logit, probit and complementary log-log, will each be considered. Examples will be given of constructing designs for the cases of $m = 1$, $m = 2$ and $m > 2$ explanatory variables. It will be shown how to obtain a locally D-optimal design for a specified parameter set if the locally D-optimal design for a related parameter set is already known. Obtaining exact designs (see Sub-section 3.3.1) will be discussed. Obtaining a design when the total number of observations is small will be considered. Lastly, there will be brief discussion on obtaining a design that is optimal when there is uncertainty about the appropriate link function to use, or which predictor variables to use, or the form of the linear predictor.

Only D- and D_S-optimality, and a new form (IMSE-optimality) are considered in this chapter.

4.2 Notation

As always, the optimal experimental design will have s support points, represented by the $m \times 1$ vectors $\boldsymbol{x}_1, \ldots, \boldsymbol{x}_s$. In the binomial situation considered in this chapter, at the ith support point n_i independent observations are taken on a binary response variable, Y_i, that has a probability of "success" given by π_i. When each n_i is equal to 1, this special case of the binomial distribution is called the Bernoulli distribution. So each individual observation has a Bernoulli distribution.

It is a standard statistical result for the Bernoulli distribution that $\mu_i = \mathrm{E}(Y_i) = \pi_i$ and $\mathrm{var}(Y_i) = \pi_i(1 - \pi_i)$, so the relationship $\mathrm{var}(Y_i) = \phi V(\mu_i)$ implies that $\phi = 1$ and $V(\mu) = \mu(1 - \mu)$.

We model a function of π_i in terms of the explanatory variables in $\boldsymbol{x}_i$ by means of a linear combination of parameters; that is, we have

$$g(\pi_i) = \eta_i = \boldsymbol{f}^\top(\boldsymbol{x}_i)\boldsymbol{\beta} \quad (i = 1, \ldots, s).$$

The aim is to select

(i) the support points from a set, $\mathcal{X} \in \mathbb{R}^m$, of possible points, and

(ii) the associated design weights $\delta_1, \ldots, \delta_s$,

so that the design

$$\xi = \left\{ \begin{array}{cccc} \boldsymbol{x}_1 & \boldsymbol{x}_2 & \ldots & \boldsymbol{x}_s \\ \delta_1 & \delta_2 & \ldots & \delta_s \end{array} \right\} \tag{4.1}$$

is "optimal" in some way.

Three commonly used link functions and various models $\eta = \boldsymbol{f}^\top(\boldsymbol{x})\boldsymbol{\beta}$ will be considered.

4.3 Link functions

The three most commonly used link functions are the *logit, probit* and *complementary log-log* functions. Each is considered below. Atkinson, Donev, & Tobias (2007, p. 399) briefly mention two other link functions.

4.3.1 The logit link function

The logit function was introduced on page 11 as

$$g(\pi) = \ln\left(\frac{\pi}{1-\pi}\right), \quad 0 < \pi < 1.$$

The logit function is the canonical link function (see page 15) for the Bernoulli and binomial distributions.

As stated in (3.7), $g(\pi_i) = \eta_i$ is equivalent to

$$\pi_i = \frac{1}{\exp(-\eta_i) + 1}. \tag{4.2}$$

Then, from (3.8),

$$\frac{\partial \pi_i}{\partial \eta_i} = \pi_i(1 - \pi_i).$$

From (3.5), the model weight $\omega(\boldsymbol{x}_i)$ associated with the ith support point is given by

$$\omega(\boldsymbol{x}_i) = \frac{1}{\text{var}(Y_i)} \left(\frac{\partial \pi_i}{\partial \eta_i}\right)^2.$$

Hence, for the logit link,

$$\omega(\boldsymbol{x}_i) = \frac{1}{\pi_i(1-\pi_i)} \times [\pi_i(1-\pi_i)]^2 = \pi_i(1-\pi_i).$$

By use of (4.2), $\omega(\boldsymbol{x}_i)$ may be written in an alternative form as

$$\omega(\boldsymbol{x}_i) = \frac{\exp(\eta_i)}{[1 + \exp(\eta_i)]^2} \tag{4.3}$$

in a proof exactly equivalent to (3.8).

Then, for a logit link, the information matrix for the design ξ in (4.1) is

$$\begin{aligned} \boldsymbol{M}(\xi,\boldsymbol{\beta}) &= \sum_{i=1}^{s} \delta_i\, \omega(\boldsymbol{x}_i)\boldsymbol{f}(\boldsymbol{x}_i)\boldsymbol{f}^{\top}(\boldsymbol{x}_i) \\ &= \sum_{i=1}^{s} \delta_i \frac{\exp(\eta_i)}{[1+\exp(\eta_i)]^2}\boldsymbol{f}(\boldsymbol{x}_i)\boldsymbol{f}^{\top}(\boldsymbol{x}_i). \end{aligned}$$

The matrix $\boldsymbol{M}(\xi,\boldsymbol{\beta})$ clearly depends on ξ through the values of the $\boldsymbol{x}_i$ and δ_i, but it also depends on the vector of parameters, $\boldsymbol{\beta}$, through $\eta_i = \boldsymbol{f}^{\top}(\boldsymbol{x}_i)\boldsymbol{\beta}$.

4.3.2 The probit link function

First consider the distribution function of the $N(0,1)$ distribution. This function is traditionally written as

$$\Phi(z) = \Pr(Z < z) \quad (-\infty < z < \infty),$$

where Z has a $N(0,1)$ distribution. For example, as $\Pr(Z < 1.96) = 0.975$, then $\Phi(1.96) = 0.975$.

The probit function, which will be used as a link function, is the *inverse* of the $N(0,1)$ distribution function. While Φ gives a cumulative probability that corresponds to a particular point or "quantile" (e.g., the cumulative probability 0.975 corresponds to the point 1.96), its inverse, Φ^{-1}, gives the point that corresponds to a particular cumulative probability. For example, the point 1.96 corresponds to the cumulative probability 0.975: $\Phi^{-1}(0.975) = 1.96$.

Remember that an *inverse function* and a *reciprocal* are not the same: if a function f is applied to a value a and results in an answer of b, then the inverse of f, f^{-1} (if it exists), is applied to b to give an answer of a.

So the probit link function is

$$g(\pi) = \Phi^{-1}(\pi), \quad 0 < \pi < 1.$$

In R, the function Φ^{-1} is known as *qnorm*. Type `qnorm(0.975)` into R, and you will get `1.959964` (1.96 to two decimal places). The R function *pnorm* represents Φ: `pnorm(1.96)` gives `0.9750021`.

As usual, the link function is equated to the linear combination of parameters: $g(\pi_i) = \Phi^{-1}(\pi_i) = \eta_i$. To find the model weight for use in the information matrix $\boldsymbol{M}(\xi,\boldsymbol{\beta})$, the partial derivative $\partial\pi_i/\partial\eta_i$ must be calculated. This may be done by noting that $\Phi^{-1}(\pi_i) = \eta_i$ implies that

$$\pi_i = \Phi(\eta_i). \tag{4.4}$$

Now

$$\Phi(\eta_i) = P(Z < \eta_i) = \int_{-\infty}^{\eta_i} \phi(z)\, dz,$$

where $\phi(z)$ represents the probability density function of the $N(0,1)$ distribution, and the integral represents the area beneath the density curve and to the left of η_i. The R function *dnorm* gives ϕ: type `dnorm(0)` in R to get $\phi(0) = 0.3989423$.

Returning to the task in hand, we have

$$\pi_i = \Phi(\eta_i) = \int_{-\infty}^{\eta_i} \phi(z)\, dz,$$

and we wish to calculate $\partial\pi_i/\partial\eta_i$. To differentiate the RHS of this equation with respect to η_i, one uses the Fundamental Theorem of Integral calculus (which appears in any reputable calculus textbook). The answer is

$$\frac{\partial\pi_i}{\partial\eta_i} = \phi(\eta_i).$$

Hence the model weight (3.5) is given by

$$\omega(\boldsymbol{x}_i) = \frac{1}{\text{var}(Y_i)}\left(\frac{\partial\pi_i}{\partial\eta_i}\right)^2 = \frac{1}{\pi_i(1-\pi_i)}[\phi(\eta_i)]^2 = \frac{[\phi(\eta_i)]^2}{\Phi(\eta_i)[1-\Phi(\eta_i)]}. \tag{4.5}$$

So, for a probit link, the information matrix for the design ξ in (4.1) is

$$\begin{aligned} \boldsymbol{M}(\xi,\boldsymbol{\beta}) &= \sum_{i=1}^{s} \delta_i\, \omega(\boldsymbol{x}_i)\boldsymbol{f}(\boldsymbol{x}_i)\boldsymbol{f}^\top(\boldsymbol{x}_i) \\ &= \sum_{i=1}^{s} \delta_i \frac{[\phi(\eta_i)]^2}{\Phi(\eta_i)[1-\Phi(\eta_i)]}\boldsymbol{f}(\boldsymbol{x}_i)\boldsymbol{f}^\top(\boldsymbol{x}_i). \end{aligned}$$

4.3.3 The complementary log-log link function

The complementary log-log function is

$$g(\pi) = \ln[-\ln(1-\pi)], \quad 0 < \pi < 1.$$

As $g(\pi_i) = \eta_i$, then π_i is calculated from η_i using

$$\pi_i = 1 - \exp[-\exp(\eta_i)]. \tag{4.6}$$

It follows that

$$\frac{\partial\pi_i}{\partial\eta_i} = \exp(\eta_i) \times \exp[-\exp(\eta_i)],$$

and so the model weight $\omega(\boldsymbol{x}_i)$ is given by

$$\begin{aligned} \omega(\boldsymbol{x}_i) &= \frac{1}{\text{var}(Y_i)}\left(\frac{\partial\pi_i}{\partial\eta_i}\right)^2 = \frac{1}{\pi_i(1-\pi_i)}\left(\frac{\partial\pi_i}{\partial\eta_i}\right)^2 \\ &= \frac{1}{\{1-\exp[-\exp(\eta_i)]\}\exp[-\exp(\eta_i)]}\{\exp(\eta_i)\times\exp[-\exp(\eta_i)]\}^2 \\ &= \frac{\exp[2\eta_i - \exp(\eta_i)]}{1-\exp[-\exp(\eta_i)]}. \end{aligned} \tag{4.7}$$

Using a complementary log-log link, the information matrix for the design ξ in (4.1) is

$$\begin{aligned}\boldsymbol{M}(\xi,\boldsymbol{\beta}) &= \sum_{i=1}^{s}\delta_i\,\omega(\boldsymbol{x}_i)\boldsymbol{f}(\boldsymbol{x}_i)\boldsymbol{f}^\top(\boldsymbol{x}_i)\\ &= \sum_{i=1}^{s}\delta_i\,\frac{\exp[2\eta_i-\exp(\eta_i)]}{1-\exp[-\exp(\eta_i)]}\boldsymbol{f}(\boldsymbol{x}_i)\boldsymbol{f}^\top(\boldsymbol{x}_i).\end{aligned}$$

4.3.4 Comparing the three link functions

For $0<\pi<1$, the logit, probit and complementary log-log link functions map π to any real number. The functions have similar shapes, but they do differ. Figure 4.1 shows what value of π results from a value of the linear combination η under each of the three link functions. This uses (4.2), (4.4) and (4.6). Rewriting $g(\pi)=\eta$ as $\pi=g^{-1}(\eta)$, where $g^{-1}(\cdot)$ is the inverse function of $g(\cdot)$, the logit and probit links both satisfy $g^{-1}(-\eta)=1-g^{-1}(\eta)$; that is, if $\eta=a$ gives the probability of success for a trial, then $\eta=-a$ gives the probability of failure. However, this relationship does not occur if the complementary log-log link is used. This fact may influence an experimenter's choice of link function for Bernoulli or binomial data.

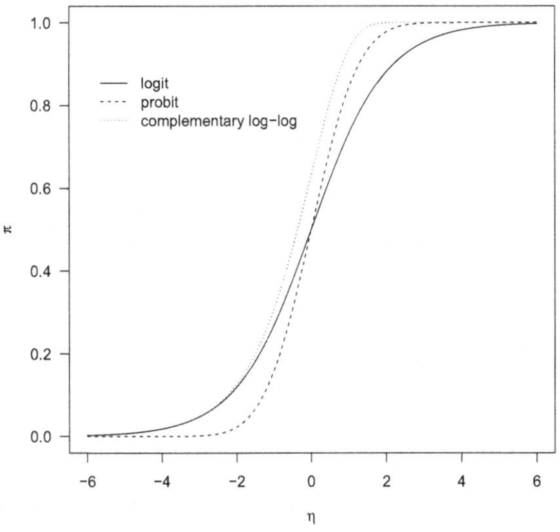

Figure 4.1 *Graphs of the relationship of π to η under each of the logit, probit and complementary log-log link functions.*

4.4 The model $\eta = \beta_0 + \beta_1 x$

4.4.1 The logit link

On page 86 it was shown that, under the logit link function

$$g(\pi) = \text{logit}(\pi) = \ln\left(\frac{\pi}{1-\pi}\right),$$

the global D-optimal design for $z = \beta_0 + \beta_1 x$ is

$$\xi^*_{\text{L},\,z} = \left\{ \begin{array}{cc} -1.5434 & 1.5434 \\ 0.5 & 0.5 \end{array} \right\}. \tag{4.8}$$

The subscript "L, z" on ξ^* indicates that this is a design for a logit link and is in terms of the canonical variable z. A design in terms of the explanatory variable x will have z replaced by x. When it is apparent to what link function and/or variable a design relates, the subscripts will be omitted.

If you find a D-optimal design for $z = \beta_0 + \beta_1 x$, D-optimal designs may be obtained for any values of β_0 and β_1. As the support points for ξ^*_z are $z = \pm 1.5434$, it follows that the support points for arbitrary values of β_0 and β_1 are found by solving $z = \beta_0 + \beta_1 x = \pm 1.5434$, which gives $x = (\pm 1.5434 - \beta_0)/\beta_1$.

That is, provided that each of $(-1.5434 - \beta_0)/\beta_1$ and $(1.5434 - \beta_0)/\beta_1$ lie in the set of acceptable values for x, the D-optimal design for the logistic model with $\eta = \beta_0 + \beta_1 x$ is

$$\xi^*_{\text{L},\,x} = \left\{ \begin{array}{cc} (-1.5434 - \beta_0)/\beta_1 & (1.5434 - \beta_0)/\beta_1 \\ 0.5 & 0.5 \end{array} \right\}. \tag{4.9}$$

This is a *locally optimal design,* as its support points depend on the particular values of β_0 and β_1 that are used.

Example 4.4.1. *Suppose that the values of β_0 and β_1 are thought to be approximately 0.55 and 1.6. To investigate how much the locally D-optimal designs vary for parameter values close to 0.55 and 1.6, one may wish to obtain locally D-optimal designs for $(\beta_0, \beta_1) = (0.6, 1.5)$ and $(\beta_0, \beta_1) = (0.5, 1.6)$. By substitution of the choices of β_0 and β_1 into (4.9), one obtains the designs ξ_1^* and ξ_2^* respectively, where*

$$\xi_1^* = \left\{ \begin{array}{cc} -1.4289 & 0.6289 \\ 0.5 & 0.5 \end{array} \right\} \quad \textit{and} \quad \xi_2^* = \left\{ \begin{array}{cc} -1.2771 & 0.6521 \\ 0.5 & 0.5 \end{array} \right\}.$$

You could search for the D-optimal designs using *constrOptim* or *optim* in R, but this is a waste of time when you can use the result in (4.9).

However, the results in (4.8) and (4.9) assume that the indicated support points lie within the sets of acceptable solutions, $\mathcal{Z}$ (say) and $\mathcal{X}$. If one or both of the points do not lie in the solution spaces, these results are of little value. It will be necessary to use constrained optimisation as described in Section 3.8.

4.4.2 The probit link

The procedure to find the globally D-optimal design for $z = \beta_0 + \beta_1 x$ for the probit link is very similar to that described for finding the globally D-optimal design for z for the logit link. See the Example in Section 3.8. As the probit model has a different model weight from the logit model, the function used to calculate $M(\xi)$ is different:

```
infodet <- function(x)
{
 info <- matrix(0,2,2)
 for (i in 1:s)
{
  pt <- x[i]
  delta <- x[i+s]
  eta <- beta0 + beta1*pt
  Phi <- pnorm(eta)
  wt <- (dnorm(eta)^2)/(Phi*(1-Phi))
  info <- info + delta*wt*matrix(c(1,pt,pt,pt^2),2,2)
 }
 -det(info)
}
```

This definition of *infodet* is similar to the definition on page 77, with the only change being that the two lines

```
  expeta <- exp(beta0 + beta1*pt)
  wt <- expeta/(1+expeta)^2
```

in the earlier program are replaced by

```
  eta <- beta0 + beta1*pt
  Phi <- pnorm(eta)
  wt <- (dnorm(eta)^2)/(Phi*(1-Phi))
```

because the probit link requires a different model weight from the logit link. Apart from that, no changes are needed to either program in Section 3.8 in order to find a globally optimal design.

When searching for an optimal design with $s = 3$ support points, the following output was obtained:

```
$par
[1] -1.13802575147546126  1.13905723115445712  1.13828201610926993
[4]  0.50000045407095395  0.00768851108184589  0.49231103455090852

$value
[1] -0.198683726130048
```

Note that z_2 and z_3 are nearly equal, and nearly the negative of z_1. Note also that $\delta_2 + \delta_3 \approx 0.5$. This suggests trying a design with $s = 2$ support points and equal weights. Doing so gives a design with support points $z = -1.138$ and $z_2 = 1.138$ approximately, and design weights of essentially 0.5. One could accept this as a satisfactory answer. Alternatively, one could investigate further and let the support points be $z = -a$ and $z = a$ with design weights $\delta_1 = \delta_2 = 0.5$. This gives the information matrix

$$\begin{aligned} \boldsymbol{M}(\xi) &= \sum_{i=1}^{2} \delta_i \, \omega(\boldsymbol{z}_i) \boldsymbol{f}(\boldsymbol{z}_i) \boldsymbol{f}^\top(\boldsymbol{z}_i) \\ &= \frac{1}{2}\omega(-a) \begin{bmatrix} 1 & -a \\ -a & a^2 \end{bmatrix} + \frac{1}{2}\omega(a) \begin{bmatrix} 1 & a \\ a & a^2 \end{bmatrix}. \end{aligned}$$

Now $\phi(-a) = \phi(a)$ (as the "bell-shaped curve" $y = \phi(x)$ is symmetric around $x = 0$) and $\Phi(-a) = 1 - \Phi(a)$; e.g., $\Phi(-1.96) = P(Z < -1.96) = P(Z > 1.96) = 1 - P(Z < 1.96) = 1 - \Phi(1.96)$. It follows that

$$\omega(-a) = \frac{[\phi(-a)]^2}{\Phi(-a)[1 - \Phi(-a)]} = \frac{[\phi(a)]^2}{[1 - \Phi(a)]\Phi(a)} = \omega(a),$$

and so

$$\begin{aligned} \boldsymbol{M}(\xi) &= \frac{1}{2}\omega(a) \begin{bmatrix} 1 & -a \\ -a & a^2 \end{bmatrix} + \frac{1}{2}\omega(a) \begin{bmatrix} 1 & a \\ a & a^2 \end{bmatrix}. \\ &= \frac{1}{2}\omega(a) \begin{bmatrix} 2 & 0 \\ 0 & 2a^2 \end{bmatrix} \\ &= \omega(a) \begin{bmatrix} 1 & 0 \\ 0 & a^2 \end{bmatrix}. \end{aligned} \tag{4.10}$$

Then the determinant of $\boldsymbol{M}(\xi)$ is a function, $h(a)$, of a, where

$$h(a) = a^2[\omega(a)]^2 = \frac{a^2[\phi(a)]^4}{\{\Phi(a)[1 - \Phi(a)]\}^2}.$$

The following program seeks the maximum of $h(a)$ (the minimum of $-h(a)$) for $a \in \{x : 1 \leq x \leq 1.5\}$.

```
detm <- function(x)
{
 num <- x^2*(dnorm(x)^4)
 Phix <- pnorm(x)
 determ <- num/(Phix*(1 - Phix))^2
 -determ
}
optimise(detm,c(1,1.5))
```

It gives the following output:

```
$minimum
[1] 1.1381


$objective
[1] -0.1986837
```

So the maximum value of $h(a)$ is 0.1986837, and occurs when $a = 1.1381$. This suggests that the globally D-optimal design is

$$\xi^* = \left\{ \begin{array}{cc} -1.1381 & 1.1381 \\ 0.5 & 0.5 \end{array} \right\}. \tag{4.11}$$

It is still necessary to verify that ξ^* is indeed D-optimal, by means of the general equivalence theorem. Substituting $a = 1.1381$ into (4.10) gives

$$\boldsymbol{M}(\xi^*) = \omega(1.1381) \left[\begin{array}{cc} 1 & 0 \\ 0 & 1.1381^2 \end{array} \right] = \left[\begin{array}{cc} 0.2087591 & 0 \\ 0 & 0.2703997 \end{array} \right].$$

The standardised variance

$$d(x, \xi^*) = \omega(x) \boldsymbol{f}^\top(x) \boldsymbol{M}^{-1}(\xi^*) \boldsymbol{f}(x) = \frac{[\phi(x)]^2}{\Phi(x)[1 - \Phi(x)]} [1, x] \boldsymbol{M}^{-1}(\xi^*) [1, x]^\top$$

can be calculated, and then plotted against x for x in the arbitrarily chosen set $\{x : -5 \leq x \leq 5\}$ using the following program:

```
optPhi <- pnorm(1.1381)
optwt <- (dnorm(1.1381)^2)/(optPhi*(1-optPhi))
infomat <- optwt*matrix(c(1,0,0,1.1381^2),2,2)
invinfomat <- solve(infomat)
stdvar <- function(x)
{
 fx <- matrix(c(1,x),2,1)
 Phi <- pnorm(x)
 wt <- (dnorm(x)^2)/(Phi*(1-Phi))
 sv <- wt*t(fx)%*%invinfomat%*%fx
 sv
}
```

Both *stdvar(-1.1381)* and *stdvar(1.1381)* are equal to 2, the value of p, and *stdvar*(x) is equal to two or less in the neighbourhood of -1.1381 and 1.1381. The graph of $d(z, \xi)$ vs. z appears in Figure 4.2. The curve peaks at $stdvar = 2$, and shows no indication of exceeding 2. Therefore, the standardised variance has achieved its maximum value of $p = 2$ at the two support points. By the general equivalence theorem, the design ξ is D-optimal.

The globally D-optimal design for $z = \beta_0 + \beta_1 x$ is

$$\xi^*_{\mathrm{P},z} = \left\{ \begin{array}{cc} -1.1381 & 1.1381 \\ 0.5 & 0.5 \end{array} \right\}. \tag{4.12}$$

It follows that, for specified values of β_0 and β_1 in the linear predictor $\eta = \beta_0 + \beta_1 x$, the locally D-optimal design for the probit link is

$$\xi^*_{\mathrm{P},x} = \left\{ \begin{array}{cc} (-1.1381 - \beta_0)/\beta_1 & (1.1381 - \beta_0)/\beta_1 \\ 0.5 & 0.5 \end{array} \right\}. \tag{4.13}$$

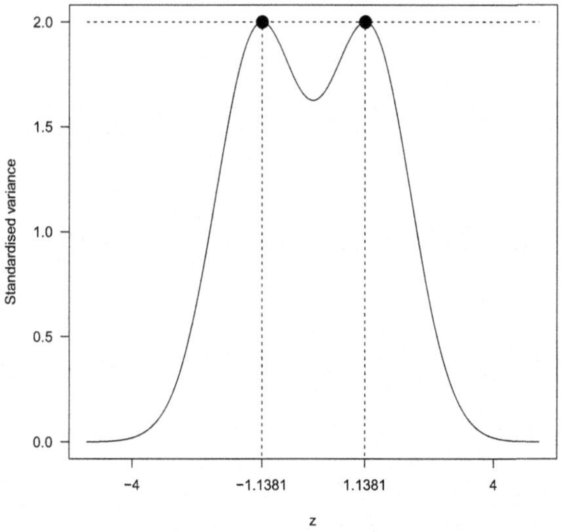

Figure 4.2 *The standardised variance, $d(z, \xi^*)$, for the design defined in* (4.11)*, plotted vs. z. Note that $d(z, \xi^*)$ attains a maximum of $p = 2$ at each of the support points.*

Example 4.4.2. *For the linear predictor $\eta = 1 + 2x$, where $\beta_0 = 1$ and $\beta_1 = 2$, the locally D-optimal design is*

$$\xi^*_{P,x} = \left\{ \begin{array}{cc} -1.06905 & 0.06905 \\ 0.5 & 0.5 \end{array} \right\}.$$

4.4.3 The complementary log-log link

As with the other two link functions, one can search for a globally D-optimal design for the complementary log-log function using the canonical variable $z = \beta_0 + \beta_1 x$. The program that was used is similar to the one used for the other two links, except that the calculation of the weight is different to take account of the model weight $\omega(z)$ given in (4.7). The commands

```
Phi <- pnorm(eta)
wt <- (dnorm(eta)^2)/(Phi*(1-Phi))
```

in the function *infodet* on page 95 are replaced by

```
expeta <- exp(eta)
wt <- exp(2*eta-expeta)/(1-exp(-expeta))
```

The constraints used earlier with *constrOptim* to search for a design with $s = 3$ support points were again used, resulting in the following output:

```
$par
[1] -1.32143e+00 7.14329e-01 9.83928e-01 4.98706e-01 9.53789e-09
[6]  5.01294e-01
$value
[1] -0.1637711
```

Recall that the output from *$par* gives $x_1, x_2, x_3, \delta_1, \delta_2, \delta_3$. The value (`9.53789e-09`) of the design weight δ_2 is negligible, suggesting that the D-optimal design requires only two support points. The other two design weights are essentially 0.5 each. So we wish to change the value of s to 2, and to give input that approximates $x_1 = -1.321$ and $x_2 = 0.984$ and $\delta_1 = \delta_2 = 0.5$. Note, however, that the input to *optim* requires values of $z_1, \ldots, z_4$ and constraints that will ensure that the x-values and δ-values satisfy the required constraints.

Here I plan to use the design region $\{x : -5 \leq x \leq 5\}$. Setting

```
zvalues1 <- acos(c(-1.321,0.984)/5)/pi
zvalues2 <- c(0.5,0.5)
zvalues2 <- sqrt(zvalues/zvalues[s])
initial <- c(zvalues1,zvalues2)
```

would give a vector *initial* that could be input to the function to be minimised. To ensure that the values used in the function satisfy the constraints $-5 \leq x_i \leq 5$; $\delta_i > 0$ $(i = 1, 2)$ and $\delta_1 + \delta_2 = 1$, the following commands are needed at the beginning of the function:

```
x <- 5*cos(pi*initial[1:s])
temp <- initial[(s+1):(2*s)]
temp2 <- temp^2
delta <- temp2/sum(temp2)
```

These changes would provide one run of *optim*, using the output from the earlier run (for $s = 3$) but now restricting the search to two support points. However, it has already been mentioned that different starting points for a search may lead to different local minima. I favour performing a number of searches, using slight variations of the initial vector, in the hope that this will have more success in finding the overall minimum. So I prefer to replace the line `initial <- c(zvalues1,zvalues2)` by something like

```
initial <- c(zvalues1,zvalues2) + 0.1*(runif(2*s)-0.5)
```

which will add small increments between -0.05 and 0.05 to each result from the previous run of the program. The overall program for 100 simulations now is the following:

①

```
f <- function(xvec)
{
 c(1,xvec)
```

```
}

infodet <- function(initial)
{
 xvec <- 5*cos(pi*initial[1:s])
 temp <- initial[(s+1):(2*s)]
 temp2 <- temp^2
 deltavec <- temp2/sum(temp2)
 info <- matrix(0,p,p)
 for (i in 1:s)
{
  fx <- f(xvec[i])
  delta <- deltavec[i]
  eta <- sum(fx*betavec)
  expeta <- exp(eta)
  wt <- exp(2*eta-expeta)/(1-exp(-expeta))
  info <- info + deltavec[i]*wt*fx%*%t(fx)
 }
 -det(info)
}
```

②

```
s <- 2
m <- 1
p <- 2
betavec <- c(0,1)
lim1 <- m*s

zvalues1 <- acos(c(-1.321,0.984)/5)/pi
zvalues2 <- c(0.5,0.5)
zvalues2 <- sqrt(zvalues2/zvalues2[s])

nsimulations <- 100
mindet <- 100
for (i in 1:nsimulations)
{
 initial <- c(zvalues1,zvalues2) + 0.1*(runif(2*s)-0.5)
 out <- optim(initial,infodet,NULL,method="Nelder-Mead")
   if(out$value < mindet) {mindet <- out$value
   bestdesign <- out$par}
 }
answer <- bestdesign
ansa <- matrix(5*cos(pi*answer[1:lim1]),m,s,byrow=T)
zvec <- answer[(lim1+1):((m+1)*s)]
deswts <- zvec^2
deswts <- deswts/sum(deswts)
solution <- rbind(ansa,deswts)
mindet
```

```
solution
```

This program will form a rough template for the programs that follow, so it is worth explaining in detail how it works. Segment ① defines $f(\boldsymbol{x})$ and calculates the negative of the determinant of the information matrix; i.e., $-\det[\boldsymbol{M}(\xi, \boldsymbol{\beta})]$. The beginning of the second function includes the transformation necessary to ensure that the values of $\boldsymbol{x}$ and $\boldsymbol{\delta}$ satisfy the necessary constraints.

In Segment ②, the values of s, m, p and $\boldsymbol{\beta}$ are specified. This is followed by the reverse transformations of the output from the previous search for an optimal design. Then the number of searches (with separate simulations of the initial values) is specified, and a large value for the minimum determinant is given. Then the optimisation is carried out repeatedly, each time using *optim*. At the end of each iteration, the design giving the minimum value of the determinant so far is recorded. At the completion of the simulations, the best design and the value of its determinant are printed out, after the arguments of the function have been converted to values satisfying the constraints.

This gave the following output:

```
> mindet
[1] -0.1637832
> solution
              [,1]      [,2]
        -1.3377372 0.9796459
deswts   0.5000002 0.4999998
```

The design weights are very close to 0.5 each. However, note that the support points are not the negative of one another, unlike in the designs for the logit and probit links. This is not surprising, as the complementary log-log link does not have the properties that $\omega(-a) = \omega(a)$ and $\hat{\pi}(-a) + \hat{\pi}(a) = 1$, where $\hat{\pi}(a)$ is the predicted value of π at $z = a$. Without these properties, it is not possible to write $\det[\boldsymbol{M}(\xi)]$ as a function of one variable in a bid to fine-tune the search for the most accurate values of the two support points. Instead, the two support points were specified to be $z = a$ and $z = b$, and the value of $\det[\boldsymbol{M}(\xi)]$ was calculated for each (a, b) combination on a grid around $(-1.3379, 0.9797)$ using the following program.

```
beta0 <- 0
beta1 <- 1

infodet2 <- function(x)
{
 info <- matrix(0,2,2)
 for (i in 1:2)
{
  pt <- x[i]
  eta <- beta0 + beta1*pt
  expeta <- exp(eta)
  wt <- exp(2*eta-expeta)/(1-exp(-expeta))
  info <- info + wt*matrix(c(1,pt,pt,pt^2),2,2)
```

```
 }
 info <- info/2
 det <- info[1,1]*info[2,2]-info[1,2]^2
 det
}
maxi <- 0
for (a in seq(from=-1.3387,to=-1.3367,by=0.0001))
{
 for (b in seq(from=0.9787,to=0.9807,by=0.0001))
 {
   detm <- infodet2(c(a,b))
   if(detm > maxi) {maxi <- detm
     soln <- c(a,b)}
 }
}
maxi
soln
```

Note that the function *infodet2* is very similar to the version of *infodet* described on page 100, except that (i) I am calculating the determinant directly using Result 2.2.2, and (ii) the determinant (rather than its negative) is calculated, as there is no use of *constrOptim* or *optim* since minimisation is not the aim. The program gave the following output:

```
> maxi
[1] 0.1637832
> soln
[1] -1.3378  0.9796
```

The output suggests that the globally D-optimal design for $z = \beta_0 + \beta_1 x$ is

$$\xi_1 = \left\{ \begin{array}{cc} -1.3378 & 0.9796 \\ 0.5 & 0.5 \end{array} \right\}. \tag{4.14}$$

To verify that ξ_1 is indeed D-optimal requires the general equivalence theorem. Using $\boldsymbol{f}(z) = (1, z)^\top$ and the formula for $\omega(\boldsymbol{z}_i)$ in (4.7), first calculate

$$\begin{aligned} \boldsymbol{M}(\xi_1) &= 0.5\times\omega(-1.3378)\boldsymbol{f}(-1.3378)\boldsymbol{f}^\top(-1.3378) \\ &\quad + 0.5\times\omega(0.9796)\boldsymbol{f}(0.9796)\boldsymbol{f}^\top(0.9976) \\ &= \left[\begin{array}{cc} 0.3805293 & 0.1068520 \\ 0.1068520 & 0.4604127 \end{array} \right] \end{aligned}$$

and then use

$$d(z, \xi_1) = \omega(z)\boldsymbol{f}^\top(z)\boldsymbol{M}^{-1}(\xi_1)\boldsymbol{f}(z)$$

to calculate the standardised variance. It equals 2 at the two support points and does not exceed 2 in the vicinity of these two points. Figure 4.3 shows $d(z, \xi_1)$ vs. z for $-5 \leq z \leq 5$, and demonstrates that $d(z, \xi_1)$ achieves its

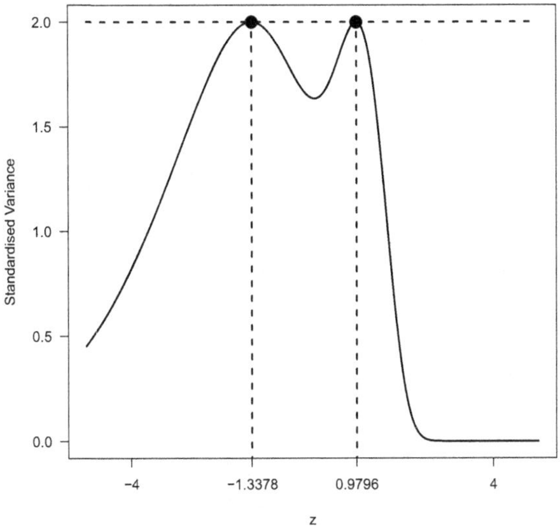

Figure 4.3 *The standardised variance $d(z, \xi_1)$ vs z for the design ξ_1 given in (4.14). Note that $d(z, \xi_1)$ attains a maximum of $p = 2$ at each of the support points.*

maximum value of $p = 2$ at the two support points of ξ_1. By the general equivalence theorem, it may be concluded that

$$\xi^*_{C,z} = \left\{ \begin{array}{cc} -1.3378 & 0.9796 \\ 0.5 & 0.5 \end{array} \right\} \tag{4.15}$$

is the globally D-optimal design for the binomial distribution when using the complementary log-log link and the canonical variable.

Consequently, for specified values of β_0 and β_1 in the linear predictor $\eta = \beta_0 + \beta_1 x$, the locally D-optimal design for the complementary log-log link is

$$\xi^*_{C,x} = \left\{ \begin{array}{cc} (-1.3378 - \beta_0)/\beta_1 & (0.9796 - \beta_0)/\beta_1 \\ 0.5 & 0.5 \end{array} \right\}. \tag{4.16}$$

Example 4.4.3. *For the linear predictor $\eta = 1+2x$, where $\beta_0 = 1$ and $\beta_1 = 2$, the locally D-optimal design is*

$$\xi^*_{C,x} = \left\{ \begin{array}{cc} -1.1689 & -0.0102 \\ 0.5 & 0.5 \end{array} \right\}.$$

4.5 The model $\eta = \beta_0 + \beta_1 x_1 + \beta_2 x_2$

4.5.1 Preliminary comments

The linear model $\eta = \beta_0 + \beta_1 x + \beta_2 x_2 = \boldsymbol{f}^\top(\boldsymbol{x})\boldsymbol{\beta}$ may be transformed to the canonical form $\eta = z_1 + z_2$ by defining $z_1 = \beta_0 + \beta_1 x_1$ and $z_2 = \beta_2 x_2$. This is equivalent to

$$\begin{bmatrix} 1 \\ z_1 \\ z_2 \end{bmatrix} = \begin{bmatrix} 1 & 0 & 0 \\ \beta_0 & \beta_1 & 0 \\ 0 & 0 & \beta_2 \end{bmatrix} \begin{bmatrix} 1 \\ x_1 \\ x_2 \end{bmatrix},$$

or

$$\boldsymbol{f}(\boldsymbol{z}) = \boldsymbol{B}\boldsymbol{f}(\boldsymbol{x}).$$

While globally D-optimal designs were successfully found in Section 4.4 using the canonical form of the linear predictor, this cannot be achieved for the case of two or more explanatory variables. Although no constraints were placed on the value of z in Section 4.4 for the logit link, a constraint effectively occurred through the model weight $\omega = \pi(1-\pi)$ becoming very small as π became close to 0 or 1. Similarly, constraints occurred through the behaviour of the model weights for the probit and complementary log-log links. However, in the (z_1, z_2) plane considered in this section, the entire length of the line $\eta = z_1 + z_2 = 0$ gives $\pi = 0.5$ for the logit and probit links, and $\pi = 1 - \exp(-1) = 0.6321$ for the complementary log-log link (from (4.2), (4.4) and (4.6), respectively). The model weight is constant along the entire line, so no constraining occurs.

There is nothing "special" about the line $z_1 + z_2 = 0$. For a given value of c, the model weight $\omega(\boldsymbol{z})$ takes a constant value along the line $\eta = z_1 + z_2 = c$. So in order to restrict the values of the explanatory variables x_1 and x_2 to "reasonable" values, constraints must be placed on them.

It has become standard to investigate locally optimal designs for values of x_1 and x_2 in the region $\mathcal{X} = \{(x_1, x_2) : -1 \leq x_1 \leq 1, -1 \leq x_2 \leq 1\}$. This is not really restrictive. If an explanatory variable, w, lies between a and $b\,(> a)$, the transformation

$$x = \frac{2w - (a+b)}{b-a} \tag{4.17}$$

gives an explanatory variable x that lies between -1 and 1. Rewriting the linear predictor η in terms of variables x_i instead of variables w_i necessitates a change to the parameter vector $\boldsymbol{\beta}$. Once the optimal design has been determined in terms of x, transform back to w using

$$w = \frac{1}{2}[(a+b) + (b-a)x].$$

This is illustrated in Example 4.5.5.

4.5.2 The logit link

Consider a situation where one seeks a locally D-optimal design on the region $\mathcal{X} = \{(x_1, x_2) : -1 \leq x_1 \leq 1, -1 \leq x_2 \leq 1\}$ using the logit link and the parameter vector $\boldsymbol{\beta} = (1, 1, 1)^\top$. Remember that this is the *assumed* parameter vector. One purpose of using the design is to estimate the actual value of $\boldsymbol{\beta}$.

The method is a straightforward extension of that demonstrated earlier in Sub-section 4.4.1. Note first that there are now $p = 3$ parameters, so that the required number of support points, s, will lie between $p = 3$ and $p(p+1)/2 = 6$. The values of x_1 and x_2 are constrained to lie between -1 and 1, and the design weights must satisfy $0 < \delta_i < 1$ $(i = 1, \ldots, s)$ and $\delta_1 + \cdots + \delta_s = 1$. If *constrOptim* is used, these constraints must be written in the form $\boldsymbol{Cv} - \boldsymbol{u} \geq \mathbf{0}$; see page 34. Note here that there are two constraints on each of the s values of x_1, two constraints on each of the s values of x_2, two constraints on each of the s individual values of δ_i, and then one remaining constraint $\delta_1 + \cdots + \delta_s = 1$; i.e., a total of $6s + 1$ constraints. Denote by x_{ij} the value of x_j at the ith support point $(i = 1, \ldots, s;\ j = 1, 2)$.

Let $\boldsymbol{x}_1 = (x_{11}, x_{21}, \ldots, x_{s1})^\top$, $\boldsymbol{x}_2 = (x_{12}, x_{22}, \ldots, x_{s2})^\top$ and $\boldsymbol{\delta} = (\delta_1, \ldots, \delta_s)^\top$. Then the constraints are written as

$$\begin{bmatrix} \boldsymbol{I}_s & \mathbf{0}_{s\times s} & \mathbf{0}_{s\times s} \\ -\boldsymbol{I}_s & \mathbf{0}_{s\times s} & \mathbf{0}_{s\times s} \\ \cdots & \cdots & \cdots \\ \mathbf{0}_{s\times s} & \boldsymbol{I}_s & \mathbf{0}_{s\times s} \\ \mathbf{0}_{s\times s} & -\boldsymbol{I}_s & \mathbf{0}_{s\times s} \\ \cdots & \cdots & \cdots \\ \mathbf{0}_{s\times s} & \mathbf{0}_{s\times s} & \boldsymbol{I}_s \\ \mathbf{0}_{s\times s} & \mathbf{0}_{s\times s} & -\boldsymbol{I}_s \\ \cdots & \cdots & \cdots \\ \mathbf{0}_s^\top & \mathbf{0}_s^\top & -\mathbf{1}_s^\top \end{bmatrix} \begin{bmatrix} \boldsymbol{x}_1 \\ \boldsymbol{x}_2 \\ \boldsymbol{\delta} \end{bmatrix} - \begin{bmatrix} -\mathbf{1}_s \\ -\mathbf{1}_s \\ \cdots \\ -\mathbf{1}_s \\ -\mathbf{1}_s \\ \cdots \\ \mathbf{0}_s \\ -\mathbf{1}_s \\ \cdots \\ -1 \end{bmatrix} \geq \begin{bmatrix} \mathbf{0}_s \\ \mathbf{0}_s \\ \cdots \\ \mathbf{0}_s \\ \mathbf{0}_s \\ \cdots \\ \mathbf{0}_s \\ \mathbf{0}_s \\ \cdots \\ 0 \end{bmatrix}.$$

I elected to start with six support points, and chose $(0.9, 0.9)^\top$, $(0.9, 0)^\top$, $(0.9, -0.9)^\top$, $(-0.9, 0.9)^\top$, $(-0.9, 0)^\top$ and $(-0.9, -0.9)^\top$. I gave each of these support points a design weight of 0.16. (Remember that the parameter values that are input to *constrOptim* must lie *inside* the region of possible values, so one must not input weights that add exactly to 1.) The input values consist of all values of x_1, followed by all values of x_2, followed by the design weights, so the following commands were used to run *constrOptim* and obtain its output.

```
initial <- c(rep(c(0.9,-0.9),each=3),rep(c(0.9,0,-0.9),2),rep(0.16,6))
out <- constrOptim(initial,detinfomat,NULL,cmat,uvec,method="Nelder-Mead")
out
```

Excerpts from the output appear below:

```
$par
 [1]  9.737512e-01  9.999900e-01  1.000000e+00 -9.340752e-01 -9.494688e-01
 [6] -9.782332e-01  7.396381e-01  1.589130e-01 -9.996839e-01  1.000000e+00
[11]  9.999991e-01 -9.988137e-01  4.828624e-05  1.030472e-01  2.993903e-01
[16]  2.584276e-01  1.394047e-01  1.996819e-01

$value
[1] -0.003549301
```

x_1:	0.9738	1.0000	1.0000	−0.9341	−0.9495	−0.9782
x_2:	0.7397	0.1589	−0.9997	1.0000	1.0000	−0.9988
δ:	0.0000	0.1030	0.2994	0.2584	0.1394	0.1997

Table 4.1 *The design suggested by the first use of* constrOptim.

```
$counts
function gradient
   10014       NA
```

That is, written to four decimal places, the values of x_1, x_2 and δ at the six support points appear in Table 4.1.

The design weight of the first point is 4.828624e-05, which is essentially numerical noise. This suggests to me that I should delete one support point and try again with five support points.

It is tedious to change the matrix $\boldsymbol{C}$ and the vector $\boldsymbol{u}$ in the constraints $\boldsymbol{Cv} - \boldsymbol{u} \geq \boldsymbol{0}$ every time that the value of s is altered. As *constrOptim* is usually slower than *optim*, and it is easier to modify input to *optim* when the value of s is changed, I will use *optim* in all future examples. I recommend that you use *optim*, too.

The following program was run. Segment ① defines the function $\boldsymbol{f}(\boldsymbol{x})$, while Segment ② defines a function, *detinfomat1*, that calculates $-\det[\boldsymbol{M}(\xi, \boldsymbol{\beta})]$. Note that the first four lines of *detinfomat1* take the input vector *variables*, convert its first ms elements into an $m \times s$ matrix whose columns are the s vectors $\boldsymbol{x}_1, \ldots, \boldsymbol{x}_s$, and convert its last s values into $\delta_1, \ldots, \delta_s$. The vectors $\boldsymbol{x}_i$ and design weights δ_i satisfy the relevant constraints. It is clear that the next command and the loop calculate

$$\boldsymbol{M}(\xi, \boldsymbol{\beta}) = \sum_{i=1}^{s} \delta_i \, \omega(\boldsymbol{x}_i) \boldsymbol{f}(\boldsymbol{x}_i) \boldsymbol{f}^\top(\boldsymbol{x}_i)$$

exactly how this formula for $\boldsymbol{M}(\xi, \boldsymbol{\beta})$ suggests that it should be calculated. The last command of the function produces the value of $-\det[\boldsymbol{M}(\xi, \boldsymbol{\beta})]$.

①

```
fx <- function(xvec)
{
 fvec <- c(1,xvec)
 fvec
}
```

②

```
detinfomat1 <- function(variables)
{
 xvals <- matrix(cos(pi*variables[1:lim1]),m,s,byrow=T)
 zvec <- variables[(lim1+1):((m+1)*s)]
 deswts <- zvec^2
 deswts <- deswts/sum(deswts)
 infomat <- matrix(0,p,p)
```

```
 for (i in 1:s)
 {
  xvec <- xvals[,i]
  fvec <- fx(xvec)
  eta <- sum(fvec*betavec)
  expeta <- exp(eta)
  wt <- expeta/((1+expeta)^2)
  infomat <- infomat + deswts[i]*wt*fvec%*%t(fvec)
 }
 -det(infomat)
}
```

The function *detinfomat1* would be used many times in searching for a locally D-optimal design. It will be worthwhile to make this function as efficient as follows, by removing the loop. The following alternative is used instead:

```
detinfomat2 <- function(variables)
{
 xvals <- cos(pi*variables[1:lim1])
 xmat <- t(matrix(xvals,s,m))
 wtvals <- (variables[(lim1+1):((m+1)*s)])^2
 deltavec <- wtvals/sum(wtvals)
 fxmat <- apply(xmat,2,fx)
 etavec <- as.vector(t(betavec)%*%fxmat)
 expetavec <- exp(etavec)
 modelwtvec <- expetavec/((1+expetavec)^2)
 infomat <- fxmat %*%diag(deltavec*modelwtvec)%*%t(fxmat)
 -det(infomat)
}
```

The following changes have been made. The matrix *xmat* in the second line is equal to $[\boldsymbol{x}_1, \ldots, \boldsymbol{x}_s]$. Then *deltavec* uses *wtvals* to produce the vector of design weights $(\delta_1, \ldots, \delta_s)^\top$. The command *apply(xmat,2,fx)* applies *fx* to each column of *xmat*. (The columns are the second dimension of the matrix: hence the "2" after *xmat*.) That is, from page 16 *fxmat* is equal to

$$\boldsymbol{F}^\top = [\boldsymbol{f}(\boldsymbol{x}_1), \ldots, \boldsymbol{f}(\boldsymbol{x}_s)].$$

The next three commands use element-by-element commands to produce the vector of model weights $\boldsymbol{\omega} = (\omega(\boldsymbol{x}_1), \ldots, \omega(\boldsymbol{x}_s))^\top$, and the following command calculates the information matrix as

$$\boldsymbol{M}(\xi, \boldsymbol{\beta}) = \boldsymbol{F}^\top \boldsymbol{W} \boldsymbol{F},$$

where $\boldsymbol{W} = \mathrm{diag}\,[\delta_1 \omega(\boldsymbol{x}_1), \ldots, \delta_s \omega(\boldsymbol{x}_s)]$.

The second form of the function, *detinfomat2,* is faster than the first form, *detinfomat1.* I have not given time comparisons here, because they depend very much on the speed of your computer and the value of s, but I was sufficiently satisfied by a comparison that I decided to use *detinfomat2* from now on.

Segment ③ (below) of the program defines $\boldsymbol{\beta}$ and the values of s, p and m. It also says how many simulated starting vectors for *optim* will be produced, and gives an initial minimum value of the negative of the determinant. It is hoped that the simulations will produce designs that provide lower values of this negative. The last part of Segment ③ produces a vector, *start*, of starting values of x and of w that *logdetinfo* will transform to constrained values of $\boldsymbol{x}_i$ and δ_i. These starting values have been obtained by "back-transforming" the last five columns of the solution from the previous search as given in Table 4.1.

③

```
betavec <- c(1,1,1)
s <- 5
p <- 3
m <- 2
lim1 <- m*s
lim2 <- (m+1)*s

nsims <- 1000
mindet <- 10

startx <- acos(c(1,1,-0.934,-0.946,-0.978,0.159,-1,1,1,-1))/pi
startw <- c(0.103,0.299,0.258,0.139,0.200)
startw <- sqrt(startw/startw[s])
start <- c(startx,startw)
```

Segment ④, the last part of the program, follows. Within a loop, each component of *start* has a small positive or negative increment added to it, then the result is input into *optim*. The resulting 'optimal' design is compared with the best found so far, and details are updated if the new design has a smaller value of $-\det[\boldsymbol{M}(\xi, \boldsymbol{\beta})]$.

④

```
for (i in 1:nsims)
{
 initial <- start + 0.1*(runif(lim2)-0.5)
 out <- optim(initial,detinfomat2,NULL,method="Nelder-Mead")
 if(out$value < mindet) {mindet <- out$value
    bestdesign <- out$par}
 }
answer <- bestdesign
xvals <- matrix(cos(pi*answer[1:lim1]),m,s,byrow=T)
zvec <- (answer[(lim1+1):((m+1)*s)])^2
deswts <- zvec/sum(zvec)
solution <- rbind(xvals,deswts)
mindet
solution
```

This gave the output

```
> mindet
[1] -0.004503855
> solution
                [,1]       [,2]       [,3]          [,4]       [,5]
        5.280236e-01  0.9999981 -0.9999998 -9.946491e-01 -0.9999995
       -6.774064e-01 -0.9999987  0.9999998  9.667211e-01 -1.0000000
```

```
deswts  1.179132e-10  0.3335182  0.3334539  8.777945e-06  0.3330191
```

on one run of the program, and

```
> mindet
[1] -0.004503852
> solution
                  [,1]       [,2]       [,3]        [,4]       [,5]
          9.941936e-01  0.9999992 -0.9999990 -0.99999871 -1.0000000
          2.090103e-01 -0.9999988  0.9999973  0.99999774 -0.9999996
deswts 1.583349e-09  0.3330912  0.2905551  0.04267929  0.3336744
```

on another run. Different results will be obtained from different runs of the program because the input for *optim* is randomly generated. Both results can be seen to suggest that the optimal design is

$$\xi^* = \left\{ \begin{array}{ccc} (1,-1)^\top & (-1,1)^\top & (-1,-1)^\top \\ 1/3 & 1/3 & 1/3 \end{array} \right\}. \tag{4.18}$$

Let us examine the behaviour of the standardised variance $d(\boldsymbol{x}, \xi^*)$ over the design region, $\mathcal{X}$. There are three variables to be considered: the explanatory variables x_1 and x_2, and $d(\boldsymbol{x}, \xi^*, \boldsymbol{\beta}) = d(x_1, x_2, \xi^*, \boldsymbol{\beta})$. *Contour plots* are a convenient way to display three-dimensional surfaces in two dimensions, and will be used often in this book when $m = 2$. The first contour plot appears in Figure 4.4, and will be explained in detail.

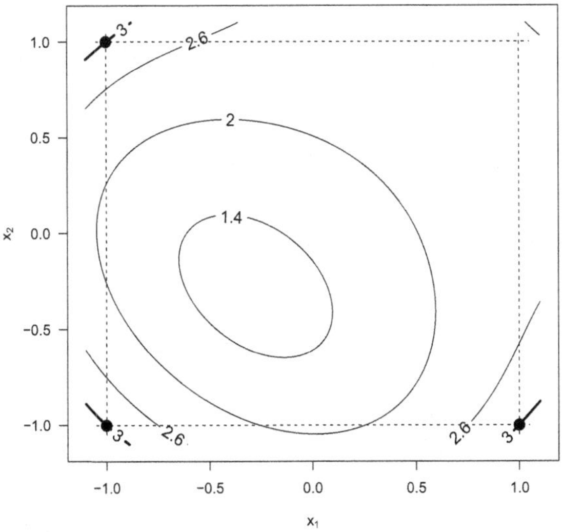

Figure 4.4 *Contour plot of $d(x_1, x_2, \xi^*, \boldsymbol{\beta})$ for $\boldsymbol{\beta} = (1, 1, 1)^\top$ over the design region $\mathcal{X} = \{-1 \leq x_1 \leq 1, -1 \leq x_2 \leq 1\}$. The design ξ^* is given in* (4.18).

The plot shows part of the x_1-x_2 plane. The dotted lines mark the boundaries of $\mathcal{X}$, and the "bullets" (•) represent the support points of ξ^*. Each curve ("contour") connects points (x_1, x_2) in $\mathcal{X}$ that have the same value of $d(x_1, x_2, \xi^*, \boldsymbol{\beta})$. For example, the curve marked "2.6" connects points (x_1, x_2) satisfying $d(x_1, x_2, \xi^*, \boldsymbol{\beta}) = 2.6$. Contours have been drawn for four values of $d(x_1, x_2, \xi^*, \boldsymbol{\beta})$: 1.4, 2, 2.6 and 3. More contours could have been drawn, but I chose to use only four in order to reduce clutter.

By the general equivalence theorem (page 83), ξ^* is locally D-optimal if the contour $d(x_1, x_2, \xi^*) = p$ (where here $p = 3$) passes through each of the support points, and there is nowhere in $\mathcal{X}$ where $d(x_1, x_2, \xi^*, \boldsymbol{\beta}) > p$. In Figure 4.4, the contour $d(x_1, x_2, \xi^*, \boldsymbol{\beta}) = 3$ is drawn with a thicker line. It is clear that this contour passes through each of the three support points, and that nowhere in $\mathcal{X}$ is $d(x_1, x_2, \xi^*, \boldsymbol{\beta}) > 3$. Hence ξ^* is locally D-optimal.

Contrast this with the contour plot in Figure 4.5, which is for the alternative design

$$\xi^+ = \left\{ \begin{array}{ccc} (1, -0.8)^\top & (-0.8, 1)^\top & (-1, -1)^\top \\ 1/3 & 1/3 & 1/3 \end{array} \right\}. \tag{4.19}$$

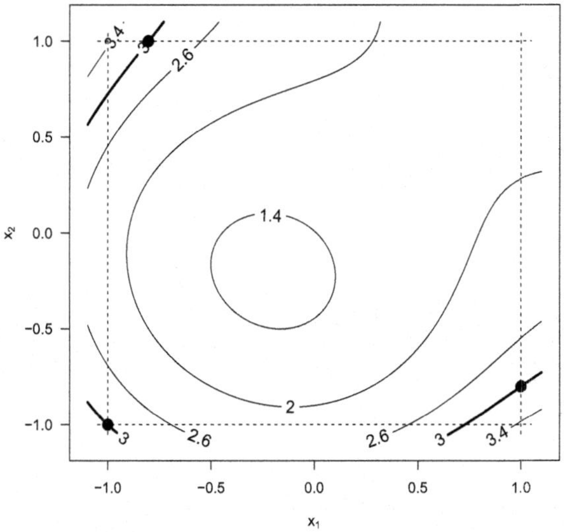

Figure 4.5 *Contour plot of* $d(x_1, x_2, \xi^+, \boldsymbol{\beta})$ *over the design region* $\mathcal{X} = \{-1 \leq x_1 \leq 1,\ -1 \leq x_2 \leq 1\}$, *where* ξ^+ *is given in* (4.19).

Contours have been drawn for $d(x_1, x_2, \xi^+, \boldsymbol{\beta}) = 1.4$, 2, 2.6, 3 and 3.4. The contour for $d(x_1, x_2, \xi^+, \boldsymbol{\beta}) = 3$ passes through the three support points of ξ^+, but the contour $d(x_1, x_2, \xi^+, \boldsymbol{\beta}) = 3.4$ also passes through $\mathcal{X}$, and it is evident that $d(x_1, x_2, \xi^+, \boldsymbol{\beta}) > 3$ for many vectors $(x_1, x_2)^\top$ in $\mathcal{X}$. Consequently, by the general equivalence theorem, the design ξ^+ is not locally D-optimal. The contour plot suggests that, if there are to be no values of $d(x_1, x_2, \xi^+, \boldsymbol{\beta})$ greater than 3 in $\mathcal{X}$, we should move the support points at $(1, -0.8)^\top$ and

$(-0.8, 1)^\top$ towards $(1, -1)^\top$ and $(-1, 1)^\top$, respectively. Of course, this makes ξ^+ more like ξ^*. It often happens that looking at the contour plot of $d(x_1, x_2, \xi, \boldsymbol{\beta})$ for a design ξ will suggest how the support points of ξ should be altered to give a design that is more D-efficient. Unfortunately, the contour plot cannot tell you anything about necessary changes to the current design weights $\delta_1, \ldots, \delta_s$.

The R program that produced the contour plot in Figure 4.4 appears below, in five segments:

①

```
m <- 2
betavec <- c(1,1,1)
p <- 3
s <- 3
lim1 <- m*s

fx <- function(xvec)
{
 fvec <- c(1,xvec)
 fvec
}
design <- c(1,-1,-1,-1,1,-1,1/3,1/3,1/3)
```

Segment ① specifies the values of m, $\boldsymbol{\beta}$, p, and s. It also calculates the value of ms (needed subsequently) and defines the function $\boldsymbol{f}(\boldsymbol{x})$. Lastly it specifies the design for which the standardised variance is being calculated: the s values of x_1 at the support points, followed by the corresponding s values of x_2, and then the corresponding design weights for these points.

②

```
 xmat <- matrix(design[1:lim1],m,s,byrow=T)
 wtvals <- (design[(lim1+1):((m+1)*s)])^2
 deltavec <- wtvals/sum(wtvals)
 fxmat <- apply(xmat,2,fx)
 etavec <- as.vector(t(betavec)%*%fxmat)
 expetavec <- exp(etavec)
 modelwtvec <- expetavec/((1+expetavec)^2)
 infomat <- fxmat %*%diag(deltavec*modelwtvec)%*%t(fxmat)
 invinfo <- solve(infomat)
```

Segment ② is based on *detinfomat2* from page 107. It calculates the information matrix $\boldsymbol{M}(\xi^*, \boldsymbol{\beta})$ for ξ^*, and then calculates $\boldsymbol{M}^{-1}(\xi^*, \boldsymbol{\beta})$. Note that the values of $\boldsymbol{x}_i$ and δ_i in *design* satisfy the necessary constraints, so they can be used directly without any transformations being required.

③

```
stdvar <- function(x1,x2)
{
 fvec <- fx(c(x1,x2))
 eta <- sum(fvec*betavec)
 expeta <- exp(eta)
 wt <- expeta/((1+expeta)^2)
```

```
 sv <- wt*as.numeric(t(fvec)%*%invinfo%*%fvec)
 sv
}
```

The function that calculates $d(x_1, x_2, \xi^*, \boldsymbol{\beta})$ is defined in Segment ③.

④

```
exp1 <- expression(x[1])
exp2 <- expression(x[2])
x1 <- seq(from=-1.1,to=1.1,by=0.005)
x2 <- seq(from=-1.1,to=1.1,by=0.005)
lenx1 <- length(x1)
lenx2 <- length(x2)
y <- rep(0,lenx1*lenx2)
kount <- 1
for (i in 1:lenx1)
{
 for (j in 1:lenx2)
 {
  y[kount] <- stdvar(x1[i],x2[j])
  kount <- kount + 1
 }
}
ymat <- matrix(y,lenx1,lenx2,byrow=T)
```

Segment ④ defines the labels that are to appear on the x_1 and x_2 axes. It then specifies the values of x_1 and x_2 to be used to create a grid of (x_1, x_2)-values at each of which the standardised variance will be calculated. The vector *y* will store these values of $d(x_1, x_2, \xi^*, \boldsymbol{\beta})$. The two loops calculate each value of $d(x_1, x_2, \xi^*, \boldsymbol{\beta})$ on the grid. The contents of *y* are then stored in a matrix for use in the command *contour*.

⑤

```
par(las=1)
contour(x1,x2,ymat,xlab=exp1,ylab=exp2, levels = c(1.4,2,2.6,3),
  lwd = c(1,1,1,3),labcex=1.2)
lines(c(-1.05,1.05),c(1,1),lty=2)
lines(c(-1.05,1.05),c(-1,-1),lty=2)
lines(c(-1,-1),c(-1.05,1.05),lty=2)
lines(c(1,1),c(-1.05,1.05),lty=2)
points(c(1,-1,-1),c(-1,1,-1),pch=16,cex=1.8)
```

In segment ⑤, the *contour* command creates the contour plot. The option *levels = c(1.4,2,2.6,3)* specifies the values of $d(x_1, x_2, \xi^*, \boldsymbol{\beta})$ for which contours are to be drawn, while *lwd = c(1,1,1,3)* specifies the widths of the contours (here the contour for $d(x_1, x_2, \xi^*, \boldsymbol{\beta}) = 3$ is to be three times the width of the other contours), and *labcex=1.2* specifies the size of the labels on the contours. The *lines* commands draw the dotted lines that mark the boundaries of the design space, while the *points* command specifies the locations, type and size of the characters that represent the support points.

As already remarked, it is clear from Figure 4.4 that $d(\boldsymbol{x}, \xi^*, \boldsymbol{\beta}) = 3 = p$ at each of the support points, and that $d(\boldsymbol{x}, \xi^*, \boldsymbol{\beta})$ does not exceed 3 anywhere on the design space.

While contour plots are very useful when there are $m = 2$ explanatory variables, their use is limited when $m > 2$. For $m = 3$ and the design region satisfying $-1 \leq x_i \leq 1$; $i = 1, 2, 3$, there would be 101 different contour plots showing the values of $d(x_1, x_2, x_3, \xi, \boldsymbol{\beta})$ for $-1 \leq x_1 \leq 1$, $-1 \leq x_2 \leq 1$ and x_3 equal to each of $-1.00, -0.98, -0.96, \ldots, 1.00$. One might also want plots of $d(x_1, x_2, x_3, \xi)$ vs (x_1, x_3) for various values of x_2, or of $d(x_1, x_2, x_3, \xi, \boldsymbol{\beta})$ vs (x_2, x_3) for various values of x_1. Producing a large number of contour plots becomes infeasible as the value of m increases.

The real aim is to see whether there are values of $d(\boldsymbol{x}, \xi, \boldsymbol{\beta})$ in the design region that exceed p. It is known from the preliminary discussion of the derivative $\phi(\boldsymbol{x}, \xi, \boldsymbol{\beta})$ on page 83 that it will be equal to zero at the support points of the design ξ. For D-optimality, this means that the standardised variance $d(\boldsymbol{x}, \xi^*, \boldsymbol{\beta})$ equals p at each support point, whether or not ξ^* is locally D-optimal. If there are values of $d(\boldsymbol{x}, \xi^*, \boldsymbol{\beta})$ greater than p in the design region, we would expect them to be close to the support points. A reasonable approach to checking that a design ξ^* is locally D-optimal is to consider values of $d(\boldsymbol{x}, \xi^*, \boldsymbol{\beta})$ in the neighbourhood of each support point, and to decide that ξ^* is locally D-optimal if the maximum value of $d(\boldsymbol{x}, \xi^*, \boldsymbol{\beta})$ in those neighbourhoods does not exceed p by an unreasonable amount. (Recall the earlier remark that, while theory says that values of $d(\boldsymbol{x}, \xi^*, \boldsymbol{\beta})$ should not exceed p at all, we must allow for computational error in performing this check.)

Assume that the function *stdvar* has been defined similarly to its definition in Segment ④ on page 85, but with the argument being $\boldsymbol{x}$ rather than x_1 and x_2 separately. For a given design ξ, the following program will examine the value of $d(\boldsymbol{x}, \xi, \boldsymbol{\beta})$ at randomly selected points in the neighbourhood of each support point of ξ. The number of points to be generated in each neighbourhood is controlled by the variable *npoints*. The distance parallel to each axis that the neighbourhood extends is equal to the parameter *distance*. The program is available in doeforglm.com as Program_13, and is as follows:

```
out4 <- matrix(c(1,-1,-1,-1,1,-1),s,m)


s <- 3
m <- 2
npoints <- 1000
total <- npoints*m
distance <- 0.02
for (i in 1:s)
{
 maxpoint <- rep(0,m)
 supportpt <- out4[i,1:m]
 max <- -1
 deviation <- 2*distance*(runif(total)-0.5)
 j1  <- 1
 j2 <- m
 for (j in 1:npoints)
 {
  newpoint <- supportpt + deviation[j1:j2]
  newpoint[newpoint < -1]  <- -1
  newpoint[newpoint > 1] <- 1
  sv <- stdvar(newpoint)
```

```
    if(sv > max) {max <- sv
      maxpoint <- newpoint}
    j1 <- j1 + m
    j2 <- j2 + m
  }
  cat("For support point ",i,"\n", "Maximum std var is ",max,
    " at \n",maxpoint,"\n")
}
```

The program follows immediately after the program that finds the allegedly D-optimal design, so that the output of the optimisation program is available to it. Running the program for the design for $\boldsymbol{\beta} = (1, 1, 1)^\top$ given on page 109 results in the output

```
For support point  1
 Maximum std var is  3  at
 1 -1
For support point  2
 Maximum std var is  3  at
 -1 1
For support point  3
 Maximum std var is  3  at
 -1 -1
```

There is no evidence that the standardised variance exceeds $p = 3$ anywhere near the alleged support points, strongly suggesting that the design is indeed locally D-optimal. Had there been values of $d(\boldsymbol{x}, \xi, \boldsymbol{\beta})$ greater than 3, we would deduce that the design is not locally D-optimal. Consider the non-optimal design ξ^+ given in (4.19), with its contour plot in Figure 4.5. Running the above program gives the output

```
For support point  1
 Maximum std var is  3.037981  at
 1 -0.8195613
For support point  2
 Maximum std var is  3.038777  at
 -0.8199656 1
For support point  3
 Maximum std var is  3  at
 -1 -1
```

While the output clearly indicates that ξ^+ is not locally D-optimal, it is not as helpful as the contour plot in indicating where the support points should be. However, it may be noticed that the maximum values of $d(\boldsymbol{x}, \xi^+, \boldsymbol{\beta})$ occur at almost the maximum distance (0.02) from the support points that is allowed by the value of *distance* in the program. This suggests that the value of *distance* should be increased. I increased it to 0.4, and the following output was obtained:

```
For support point  1
 Maximum std var is  3.433078  at
 1 -0.9966062
For support point  2
```

```
 Maximum std var is  3.441438  at
 -0.9999377 1
For support point  3
 Maximum std var is  3  at
 -1 -1
```

In the neighbourhood of the current support points, the two points where the maximum values of the standardised variance occur are vertices of the design space. As this is where the support points of optimal designs are often located, it suggests that we should try replacing the support points $(1, -0.8)$ and $(-0.8, 1)$ by $(1, -1)$ and $(-1, 1)$, respectively, and then investigate the new design (which we know from page 110 is actually the locally D-optimal design).

Atkinson, Donev, & Tobias (2007, pp. 402–410) provided D-optimal designs for four parameter sets: $\boldsymbol{\beta} = (0, 1, 1)$, $(0, 2, 2)$, $(2, 2, 2)$ and $(2.5, 2, 2)$. For the first three parameter sets, the D-optimal design has $s = 4$ support points; for the fourth parameter set, the D-optimal design has $s = 3$ support points. So clearly the number of support points depends on the values of the parameters. Moreover, for $\boldsymbol{\beta} = (0, 2, 2)$, there are two possible D-optimal designs with four support points. We shall see in Example 4.5.2 that the existence of a second D-optimal design is entirely predictable once we have found either of the D-optimal designs.

A table has been provided in the Web site doeforglm.com that gives a D-optimal design for various parameter sets satisfying $\beta_0 \in \{0, 1, 2\}$, $\beta_1 \in \{1, 2, 3, 4, 5\}$ and $\beta_2 \in \{1, \ldots, \beta_1\}$. For example, given $\boldsymbol{x} \in \mathcal{X}$, the D-optimal design for $\boldsymbol{\beta} = (1, 2, 1)$ (i.e., $\eta = \boldsymbol{f}^\top(\boldsymbol{x})\boldsymbol{\beta} = 1 + 2x_1 + x_2$) has $s = 4$ support points, and is given by

x_1	x_2	δ
−1.000	1.000	0.301
−0.761	−1.000	0.325
−0.378	1.000	0.049
0.760	−1.000	0.325

.

I have tried to give all values of the explanatory variables and design weights to three decimal places in the supplementary tables, as anything with more than three decimal places is generally impractical from an experimental perspective. Other slightly different designs could give the same value of $-\det[\boldsymbol{M}(\xi, \boldsymbol{\beta})]$. A contour plot of the standardised variance $d(x_1, x_2)$ appears in Figure 4.6. The heavy contour line for $d(x_1, x_2) = p = 3$ does not cross the dotted lines which form the boundary of the design space $\mathcal{X}$, so the maximum value of $d(x_1, x_2)$ on the boundary or interior of $\mathcal{X}$ is $p = 3$. The values of $d(x_1, x_2)$ at the tabulated support points are given in R as

```
> stdvar(-1,1)
[1] 3.001838
> stdvar(-0.761,-1)
[1] 2.99879
> stdvar(-0.378,1)
[1] 3.004101
> stdvar(0.760,-1)
[1] 2.99889
```

These do not equal 3 exactly, despite the general equivalence theorem saying that they should. You must expect this when the values of the x_i and δ_i are given to three decimal places. If you find the values of the x_i and δ_i to (say) eight decimal places, then you might justly expect the values of $d(x_1, x_2)$ at the support points to be

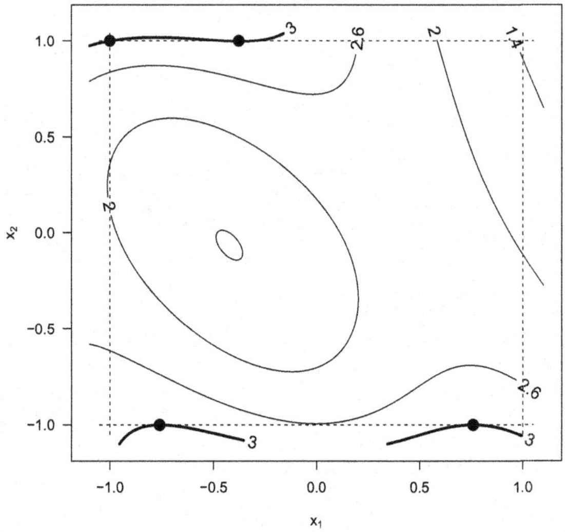

Figure 4.6 *Contour plot of the standardised variance of the locally D-optimal design for a logit link and* $\eta = 1 + 2x_1 + x_2$. *The contour labels were moved to the edges so that the two "bullets" on the line* $x_2 = -1$ *would not overlay the labels.*

much closer to 3. However, it is quite possible that they still will not equal 3 exactly. Mathematical results and computational results do not always coincide exactly.

Notwithstanding this, I strongly encourage you to check any design that you take, or modify, from the tables, to guard against errors on my part or yours. Minor differences between designs are probably the natural consequence of optimising a complicated function of numerous variables, but major differences suggest an error by you or me.

The D-optimal designs in the table are meant to provide a rough guide to choosing an initial guess of the solution for some other value of $\boldsymbol{\beta}$. For example, should you wish to find a D-optimal design for the parameters $\boldsymbol{\beta} = (1.2, 2.1, 1.3)$, you might say that this parameter set is similar to $\boldsymbol{\beta} = (1, 2, 1)$ and use the locally D-optimal design for $\boldsymbol{\beta} = (1, 2, 1)$ as your initial guess in a search for the locally D-optimal design for $\boldsymbol{\beta} = (1.2, 2.1, 1.3)$.

While this table is useful, it might seem rather limited, as it does not consider values of β_1 less than those of β_2, nor does it consider negative values of β_0, β_1 or β_2. However, the following "Results" will allow these designs to be used for values of $\boldsymbol{\beta}$ that do not meet the requirements of the parameter vectors in the table. The "Results" also apply to more than $m = 2$ mathematically independent explanatory variables, so they will be written for the more general model $\eta = \beta_0 + \beta_1 x_1 + \cdots + \beta_m x_m$. The design region is written as

$$\mathcal{X}_m = \{(x_1, x_2, \ldots, x_m) : -1 \leq x_1 \leq 1, -1 \leq x_2 \leq 1, \ldots, -1 \leq x_m \leq 1\}.$$

It is clear that the design space $\mathcal{X}$ that we have been considering is $\mathcal{X}_2$, and that the tabulated locally D-optimal designs are optimal in $\mathcal{X}_2$ for the relevant values of $\boldsymbol{\beta}$.

Let

$$\xi_{\boldsymbol{\beta}} = \left\{ \begin{array}{cccc} \boldsymbol{x}_1 & \boldsymbol{x}_2 & \ldots & \boldsymbol{x}_s \\ \delta_1 & \delta_2 & \ldots & \delta_s \end{array} \right\} \text{ and } \xi_{\boldsymbol{b}} = \left\{ \begin{array}{cccc} \boldsymbol{u}_1 & \boldsymbol{u}_2 & \ldots & \boldsymbol{u}_s \\ \delta_1 & \delta_2 & \ldots & \delta_s \end{array} \right\} \tag{4.20}$$

be designs on $\mathcal{X}_m$ for the parameter vectors $\boldsymbol{\beta} = (\beta_0, \beta_1, \ldots, \beta_m)^\top$ and $\boldsymbol{b} = (b_1, b_2, \ldots, b_m)^\top$, respectively. Note that the sets of design weights for $\xi_{\boldsymbol{\beta}}$ and $\xi_{\boldsymbol{b}}$ are equal. Let $\boldsymbol{M}(\xi_{\boldsymbol{\beta}}, \boldsymbol{\beta})$ and $\boldsymbol{M}(\xi_{\boldsymbol{b}}, \boldsymbol{b})$ be the information matrices of $\xi_{\boldsymbol{\beta}}$ and $\xi_{\boldsymbol{b}}$, respectively.

Result 4.5.1. *If $\boldsymbol{b}$ is obtained from $\boldsymbol{\beta}$ by interchanging β_j and β_k from $\{\beta_1, \ldots, \beta_m\}$ (NB: this set does not contain β_0), and each support point $\boldsymbol{u}_i$ is obtained from the corresponding $\boldsymbol{x}_i$ by interchanging the values of the explanatory variables x_j and x_k, then $det\left[\boldsymbol{M}(\xi_{\boldsymbol{\beta}}, \boldsymbol{\beta})\right] = det[\boldsymbol{M}(\xi_{\boldsymbol{b}}, \boldsymbol{b})]$.*

Result 4.5.2. *If $\boldsymbol{b}$ is obtained from $\boldsymbol{\beta}$ by changing the sign (+ or −) of $\beta_j \in \{\beta_1, \ldots, \beta_m\}$, and each $\boldsymbol{u}_i$ is obtained from the corresponding $\boldsymbol{x}_i$ by changing the sign of the explanatory variable x_j, then $det\left[\boldsymbol{M}(\xi_{\boldsymbol{\beta}}, \boldsymbol{\beta})\right] = det[\boldsymbol{M}(\xi_{\boldsymbol{b}}, \boldsymbol{b})]$.*

[Acknowledgment of priority: Since formulating Result 4.5.1, I have discovered a similar statement in Zhang (2006, pp. 50-51).]

Results 4.5.1 and 4.5.2 allow us to obtain locally D-optimal designs for parameter sets related to those for which locally D-optimal designs have already been tabulated, without doing a constrained optimisation. Note that the design space $\mathcal{X}_m$ is unchanged if we swap the order in which we write some of the coordinates $x_1, \ldots, x_m$, and if we replace some x_i by $-x_i$ (as $-1 \leq x_i \leq 1 \Leftrightarrow -1 \leq -x_i \leq 1$). For example, the linear predictor $\eta = 1 + 2x_1 + x_2$ is equal to $\eta = 1 + 2x_1 - 1(-x_2) = (1, 2, -1)(1, x_1, -x_2)^\top$, so finding the design that maximises $\boldsymbol{M}(\xi_{\boldsymbol{\beta}}, \boldsymbol{\beta})$ for $\boldsymbol{\beta} = (1, 2, 1)^\top$ over $\mathcal{X}_2$ is the same as finding the design that maximises $\boldsymbol{M}(\xi_{\boldsymbol{b}}, \boldsymbol{b})$ for $\boldsymbol{b} = (1, 2, -1)^\top$ over the set of values of $(x_1, -x_2)$ corresponding to $(x_1, x_2) \in \mathcal{X}$; that is, over $\mathcal{X}_2$.

Example 4.5.1. *From page 115, if $\boldsymbol{x} \in \mathcal{X}_2$, the locally D-optimal design for $\boldsymbol{\beta} = (1, 2, 1)$ for the logic link is given by*

x_1	x_2	δ
−1.000	1.000	0.301
−0.761	−1.000	0.325
−0.378	1.000	0.049
0.760	−1.000	0.325

Then the locally D-optimal designs for $\boldsymbol{\beta} = (1, 1, 2)$, $(1, 2, -1)$ and $(1, -1, -2)$ are

	$\boldsymbol{\beta} = (1, 1, 2)^\top$		$\boldsymbol{\beta} = (1, 2, -1)^\top$		$\boldsymbol{\beta} = (1, -1, -2)^\top$	
δ	x_1	x_2	x_1	x_2	x_1	x_2
0.301	1.000	−1.000	−1.000	−1.000	−1.000	1.000
0.325	−1.000	−0.761	−0.761	1.000	1.000	0.761
0.049	1.000	−0.378	−0.378	−1.000	−1.000	0.378
0.325	−1.000	0.760	0.760	1.000	1.000	−0.760

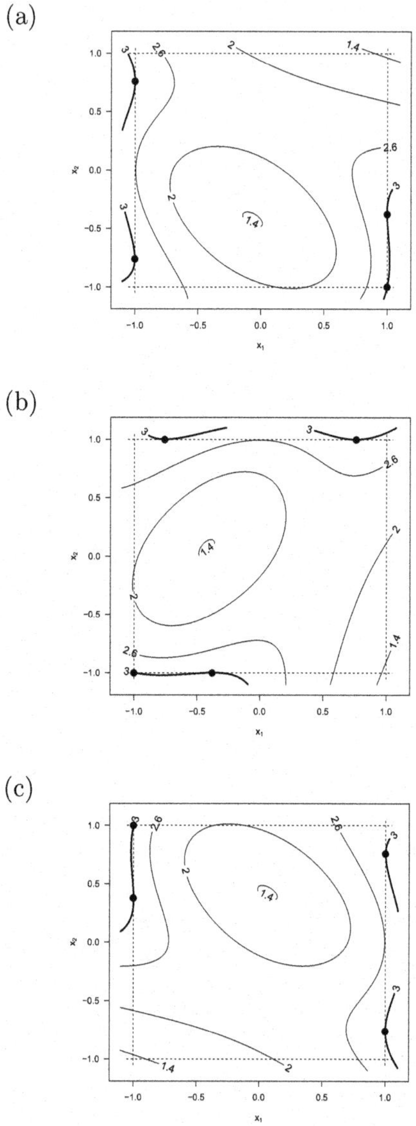

Figure 4.7 *Plots of the standardised variances of the designs given in Example 4.5.1 for the parameter vectors (a)* $(1,1,2)^\top$*, (b)* $(1,2,-1)^\top$ *and (c)* $(1,-1,-2)^\top$*. As the standardised variances for each value of* $\boldsymbol{\beta}$ *achieve their maximum value over the design region of* $p=3$ *at the marked support points, the designs are each locally D-optimal for the relevant value of* $\boldsymbol{\beta}$*.*

The contour plots of the standardised variances for these three designs appear in Figure 4.7. For each plot, the standardised variance achieves its maximum value over the design region of $p = 3$ at each of the support points. Therefore, by the general equivalence theorem, each design is locally D-optimal for the relevant value of $\boldsymbol{\beta}$.

Example 4.5.2. *From the table, the D-optimal design for $\boldsymbol{\beta} = (0, 2, 2)^\top$ is*

$$\xi_a = \left\{ \begin{array}{cccc} (-1.000, 0.118)^\top & (-1.000, 1.000)^\top & (-0.118, 1.000)^\top & (1.000, -1.000)^\top \\ 0.240 & 0.193 & 0.240 & 0.327 \end{array} \right\},$$

which agrees within numerical noise with a design given in Atkinson, Donev, & Tobias (2007, Design B2, p. 403). The authors also find a second D-optimal design,

$$\xi_b = \left\{ \begin{array}{cccc} (-1.000, 1.000)^\top & (1.000, -1.000)^\top & (1.000, -0.118)^\top & (0.118, -1.000)^\top \\ 0.327 & 0.193 & 0.240 & 0.240 \end{array} \right\}.$$

The D-optimality of either design implies the D-optimality of the other by a simple application of Result 4.5.1. Interchanging x_1 and x_2 in the linear predictor $\eta = 2x_1 + 2x_2$ gives exactly the same model, while interchanging the values of x_1 and x_2 in the support points of ξ_a gives ξ_b and vice versa.

In the discussions so far, some of the parameters $\beta_1, \ldots, \beta_m$ in the linear predictor $\eta = \beta_0 + \beta_1 x_1 + \cdots + \beta_m x_m$ have had their signs changed, or have been interchanged, to obtain D-optimal designs based on those in the online documentation. However, changing the sign of β_0 has not yet been considered. For the logit and probit links, for which $\pi(-\eta_i) = 1 - \pi(\eta_i)$, one may change the sign of β_0 and obtain a locally D-optimal design by simple modification of a tabulated design. Note that this does not apply to the complementary log-log link, as $\pi(-\eta_i) \neq 1 - \pi(\eta_i)$.

Result 4.5.3. *If the link function satisfies $\pi(-\eta_i) = 1 - \pi(\eta_i)$, then the locally D-optimal design on $\mathcal{X}_m$ for $\mathbf{b} = -\boldsymbol{\beta}$, $\xi^*_{\mathbf{b}}$, is identical to the locally D-optimal design on $\mathcal{X}_m$ for $\boldsymbol{\beta}$, $\xi^*_{\boldsymbol{\beta}}$.*

Example 4.5.3. *The locally D-optimal design on $\mathcal{X}$ for the parameter vector $\mathbf{b} = (-1, -2, -1)^\top$ is identical to the locally D-optimal design on $\mathcal{X}$ for $\boldsymbol{\beta} = (1, 2, 1)^\top$.*

Example 4.5.4. *The locally D-optimal design on $\mathcal{X}$ for the parameter vector $\mathbf{b} = (-1, 2, 1)^\top$ is identical to the locally D-optimal design on $\mathcal{X}$ for $\boldsymbol{\beta} = (1, -2, -1)^\top$. The latter design is not given in the supplementary tables, but it is known from Results 4.5.1 and 4.5.2 that it is obtained from the locally D-optimal design for $\boldsymbol{\beta} = (1, 2, 1)^\top$ (which is given in the tables) by changing the sign (+ to −, or vice versa) of the values of x_1 and x_2 at each support point of the locally D-optimal design for $\boldsymbol{\beta} = (1, 2, 1)^\top$.*

Example 4.5.5. *This example considers a situation where the predictor variables, w_1 and w_2, do not lie between −1 and 1. Consider the logit link and let the linear predictor be $\eta = -4 + 2w_1 + w_2$, where $0 < w_1 < 3$ and $1 < w_2 < 5$. From (4.17), the transformations*

$$x_1 = (w_1 - 1.5)/1.5 \quad \text{and} \quad x_2 = (w_2 - 3)/2$$

give new variables x_1 and x_2 satisfying $-1 < x_i < 1$ $(i = 1, 2)$. The inverse transformations are $w_1 = 1.5 + 1.5x_1$ and $w_2 = 3 + 2x_2$. Substituting these results into the equation for η gives

$$\begin{aligned} \eta &= -4 + 2w_1 + w_2 \\ &= -4 + 2(1.5 + 1.5x_1) + (3 + 2x_2) \\ &= 2 + 3x_1 + 2x_2, \end{aligned}$$

ξ_x			ξ_w	
x_1	x_2	δ	w_1	w_2
-1.000	-0.232	0.082	0.000	2.536
-1.000	1.000	0.326	0.000	5.000
-0.493	-1.000	0.265	0.761	1.000
0.493	-1.000	0.327	2.240	1.000

Table 4.2 *Locally D-optimal designs for the problem in Example 4.5.5. The three leftmost columns give the design in terms of the transformed explanatory variables x_1 and x_2, while the three rightmost columns give the design in terms of the original explanatory variables w_1 and w_2.*

which has x_1 and x_2 in the usual design space for a parameter vector $\boldsymbol{\beta} = (2, 3, 2)^\top$. The locally D-optimal design is obtainable from the optimal designs in the online resources. It (ξ_x) appears in the three left-hand columns of Table 4.2, while the locally D-optimal design in terms of w_1 and w_2 (ξ_w) appears in the three right-hand columns. The support points in terms of w_1 and w_2 are obtained from those of x_1 and x_2 using the inverse transformations above.

It is easy to verify that ξ_w does have a standardised variance satisfying the requirements of the general equivalence theorem, and so ξ_w is indeed locally D-optimal.

Result 4.5.3 occurs because of the following: Write $\eta_{\boldsymbol{b}} = \boldsymbol{f}^\top(\boldsymbol{x})\boldsymbol{b}$ and $\eta_{\boldsymbol{\beta}} = \boldsymbol{f}^\top(\boldsymbol{x})\boldsymbol{\beta}$. If $\boldsymbol{b} = -\boldsymbol{\beta}$, then $\eta_{\boldsymbol{b}} = -\eta_{\boldsymbol{\beta}}$ and, at a support point $\boldsymbol{x}_i$, the model weights for the two models are equal, because $\pi_{\boldsymbol{\beta}} = 1 - \pi_{\boldsymbol{b}}$ implies that

$$\omega_{\boldsymbol{\beta}}(\boldsymbol{x}_i) = \pi_{\boldsymbol{\beta}}(1 - \pi_{\boldsymbol{\beta}}) = (1 - \pi_{\boldsymbol{b}})\pi_{\boldsymbol{b}} = \omega_{\boldsymbol{b}}(\boldsymbol{x}_i).$$

Also, we may write $\eta(\boldsymbol{u}) = -\boldsymbol{f}^\top(\boldsymbol{x})\boldsymbol{\beta} = [-\boldsymbol{f}(\boldsymbol{x})]^\top\boldsymbol{\beta} = \boldsymbol{f}^\top(\boldsymbol{u})\boldsymbol{\beta}$ where $\boldsymbol{f}(\boldsymbol{u}) = -\boldsymbol{f}(\boldsymbol{x})$. Then, for an arbitrary design ξ,

$$\begin{aligned}\boldsymbol{M}(\xi, \boldsymbol{b}) &= \sum_{i=1}^{s} \delta_i\, \omega(\boldsymbol{u}_i)\boldsymbol{f}(\boldsymbol{u}_i)\boldsymbol{f}^\top(\boldsymbol{u}_i)\\ &= \sum_{i=1}^{s} \delta_i\, \omega(\boldsymbol{x}_i)[-\boldsymbol{f}(\boldsymbol{x}_i)][-\boldsymbol{f}^\top(\boldsymbol{x}_i)]\\ &= \sum_{i=1}^{s} \delta_i\, \omega(\boldsymbol{x}_i)\boldsymbol{f}(\boldsymbol{x}_i)\boldsymbol{f}^\top(\boldsymbol{x}_i)\\ &= \boldsymbol{M}(\xi, \boldsymbol{\beta}),\end{aligned}$$

and it follows that $\boldsymbol{M}(\xi, \boldsymbol{b})$ and $\boldsymbol{M}(\xi, \boldsymbol{\beta})$ both achieve their maximum determinants at the same design.

4.5.3 The probit link

You are referred back to Sub-section 4.5.1 for a quick consideration of the reasons for working with the design space $\mathcal{X} = \{(x_1, x_2)\colon\ -1 \le x_1 \le 1;\ -1 \le x_2 \le 1\}$.

The only difference between the approach described in Sub-section 4.5.2 for the logit link and the approach for the probit link is in a change for the formula for the model weights. In the program that appears on page 107, replace the two commands:

```
expetavec <- exp(etavec)
modelwtvec <- expetavec/((1+expetavec)^2)
```

in the function *detinfomat2* by the commands:

```
Phivec <- pnorm(etavec)
modelwtvec <- (dnorm(etavec)^2)/(Phivec*(1-Phivec))
```

As with the logit link, a table has been provided in the online material that gives a locally D-optimal design for the probit link for the parameter sets (β_0, β_1, β_2) that satisfy $\beta_0 \in \{0, 1, 2\}$, $\beta_1 \in \{1, 2, 3, 4, 5\}$ and $\beta_2 \in \{1, \ldots, \beta_1\}$. Results 4.5.1, 4.5.2 and 4.5.3 may all be used with this table to obtain D-optimal designs for related parameter sets. For example, the locally D-optimal design for $\boldsymbol{\beta} = (1, 3, 2)^\top$ may be trivially altered to give locally D-optimal designs for any of $\boldsymbol{\beta} = (\pm 1, \pm 3, \pm 2)^\top$ or $(\pm 1, \pm 2, \pm 3)^\top$ without further computation.

4.5.4 *The complementary log-log link*

You should re-read Sub-section 4.5.1 before proceeding.

The only necessary change in the program for the logit link that appears on page 107 is to introduce the appropriate model weight for the complementary log-log link. Replace the command

```
modelwtvec <- expetavec/((1+expetavec)^2)
```

in the function *detinfomat2* in Segment ② by the command

```
modelwtvec <- exp(2*etavec-expetavec)/(1-exp(-expetavec))
```

A table in the Web site doefoglm.com lists a locally D-optimal design for the complementary log-log link for the parameter sets (β_0, β_1, β_2) that satisfy $\beta_0 \in \{-2, -1, 0, 1, 2\}$, $\beta_1 \in \{1, 2, 3, 4, 5\}$ and $\beta_2 \in \{1, \ldots, \beta_1\}$. Results 4.5.1 and 4.5.2 may be used with this table to obtain D-optimal designs for related parameter sets. However, Result 4.5.3 cannot be used, as the complementary log-log link function does not satisfy $\pi(-\eta) = 1 - \pi(\eta)$.

4.5.5 *Comparison of designs for different links*

The locally D-optimal design for a logit link with $m = 2$ and $\boldsymbol{\beta} = (1, 2, 1)^\top$ was listed on page 115. It is repeated in Table 4.3, together with the locally D-optimal designs for the same value of $\boldsymbol{\beta}$ and the probit and complementary log-log links.

While the designs have the same number of support points, there are considerable differences between some support points and design weights for each pair of locally D-optimal designs.

Sometimes the D-optimal designs for the same value of $\boldsymbol{\beta}$ but with different links do have different numbers of support points. An example is $\boldsymbol{\beta} = (0, 5, 5)^\top$. The optimal designs for the three links appear in the online supplementary tables. The design for the logit link has five support points, while the designs for the probit and complementary log-log links have four support points each. Fairly sustained attempts to find a locally D-optimal design with four support points for the logit link have been unsuccessful. Similarly, attempts to find locally D-optimal optimal designs with

Logit			Probit			Complementary log-log		
x_1	x_2	δ	x_1	x_2	δ	x_1	x_2	δ
-1.000	1.000	0.301	-1.000	1.000	0.292	-1.000	1.000	0.179
-0.761	-1.000	0.325	-0.560	-1.000	0.322	-0.644	-1.000	0.305
-0.378	1.000	0.049	-0.523	1.000	0.064	-0.594	1.000	0.195
0.760	-1.000	0.325	0.560	-1.000	0.322	0.478	-1.000	0.321

Table 4.3 *Locally D-optimal designs when* $\boldsymbol{\beta} = (1, 2, 1)^\top$ *for the logit, probit and complementary log-log link functions.*

five support points for the probit and complementary log-log links have also been unsuccessful. The lack of commonality in the number of support points across the three links is a consequence of the different properties of those links.

4.6 Designs for $m > 2$ explanatory variables

Most books on the Design of Experiments are declaratory in nature. This sub-section will be quite different. Finding a locally D-optimal design for a GLM can sometimes require luck as well as skill, and I aim to illustrate here the various steps that I took in a long search for the locally D-optimal design for the logit link and a linear predictor that involves more explanatory variables than we have seen before.

Suppose that there are $m = 6$ explanatory variables, and the linear predictor is assumed to be

$$\begin{aligned}\eta &= 0.4 + 0.3x_1 + 0.4x_2 - 0.5x_3 - 0.2x_4 + 0.3x_5 + 0.4x_6 + 0.2x_1x_2 - 0.3x_3x_4 \\ &= \boldsymbol{f}^\top(\boldsymbol{x})\boldsymbol{\beta}.\end{aligned}$$

There are $p = 9$ parameters in the vector $\boldsymbol{\beta}$, so from Sub-section 3.7.1 the number of support points in the locally D-optimal design will be between $p = 9$ and $p(p+1)/2 = 45$. No additional theory is needed to find a locally D-optimal design for this larger number of support points, but the procedure will take longer and probably be less straightforward than in previous sections. As well, it is not feasible to draw all the possible contour plots of the standardised variance, so it becomes necessary to check the values of the standardised variance near a support point using Program_13 on page 113.

I chose to begin a search for the locally D-optimal design with $s = 15$ support points. While this is not close to the maximum of 45 points, I felt that it was sufficiently greater than the minimum of nine points to be worth an investigation. The program used was a simple adaptation of the programs described earlier, commencing with Segment ①.

①

```
betavec <- c(0.4,0.3,0.4,-0.5,-0.2,0.3,0.4,0.2,-0.3)
p <- length(betavec)
m <- 6
s <- 15
lim1 <- m*s
lim2 <- (m+1)*s

fx <- function(xvec)
```

```
{
 fvec <- c(1,xvec,xvec[1]*xvec[2],xvec[3]*xvec[4])
 fvec
}
```

The function *detinfomat2* was then defined (see page 107), and Segment ② (below) was used to search for an optimal design. The starting values of the input vector *initial* were randomly generated 2000 times, and the design with the maximum value of $\det[\boldsymbol{M}(\xi, \boldsymbol{\beta})]$ was recorded. Segment ③ will print out the selected design.

②

```
nsims <- 2000
mindet <- 100
for (i in 1:nsims)
{
 initial <- runif(lim2)
 out <- optim(initial,detinfomat2,NULL,method="Nelder-Mead")
 valuenow <- out$val
 if(valuenow < mindet){mindet <- valuenow
 design <- out$par}
}
```

③

```
cat("Min value of det\n",mindet,"\n")
output <- design
out1 <- cos(pi*output[1:lim1])
out2 <- (output[(lim1+1):lim2])^2
wts <- out2/sum(out2)
out4 <- cbind(matrix(out1,s,m),wts)
cat("Design\n")
t(out4)
```

The selected design had a value of 5.945375×10^{-10} for $\det[\boldsymbol{M}(\xi, \boldsymbol{\beta})]$. As two of the design weights, δ_{10} and δ_{11}, were less than 0.01, I decided to reduce the value of s to 13 and, with this alteration, ran the program again. For input to the program, I used the design ξ with the 10th and 11th support points deleted. Segment ② above was replaced by Segment ④ below:

④

```
spare <- out4
xout <- as.vector(out4[c(1:9,12:15),1:6])
wtout <- as.vector(out4[c(1:9,12:15),7])

betavec <- c(0.4,0.3,0.4,-0.5,-0.2,0.3,0.4,0.2,-0.3)
p <- length(betavec)
m <- 6
s <- 13
lim1 <- m*s
lim2 <- (m+1)*s
```

```
nsims <- 2000
mindet <- 100
initialx <- xout
initialz1 <- acos(initialx)/pi
initialwts <- wtout
initialz2 <- sqrt(initialwts/initialwts[s])
for (i in 1:nsims)
{
 initial1 <- initialz1 + 0.1*(runif(lim1)-0.5)
 initial2 <- initialz2 + 0.1*(runif(s)-0.5)
 initial <- c(initial1,initial2)
 out <- optim(initial,detinfomat2,NULL,method="Nelder-Mead")
 valuenow <- out$val
 if(valuenow < mindet){mindet <- valuenow
 design <- out$par}
}
```

The first command of Segment ④ (`spare <- out4`) is not needed for the search for an optimal design but, as I discovered later, it is wise to have a spare copy of the output, *out4*, from the previous run of the program.

I ran the program three times, each time using the "best" design from the previous run as a baseline from which to generate input to *optim*. This required the second and third lines of Segment ④ to be replaced by Segment ⑤:

⑤

```
xout <- as.vector(out4[,1:6])
wtout <- as.vector(out4[,7])
```

Running the program resulted in a design with $\det[\boldsymbol{M}(\xi,\boldsymbol{\beta})] = 1.564124 \times 10^{-8}$ and δ_8 less than 0.01. (All values of δ_i for the previous two designs were greater than 0.01.) The final design had the eighth support point deleted, s was changed to 12, and then the program was run again. After 15 iterations, each time with $\delta_{\text{min}} > 0.033$, the value of $\det[\boldsymbol{M}(\xi,\boldsymbol{\beta})]$ had increased to 3.299177×10^{-7}. The next three iterations gave designs with lower values of $\det[\boldsymbol{M}(\xi,\boldsymbol{\beta})]$, and this was where it was important to have stored a copy of the matrix *out4* from the best design so far. Without this, one would have to use the less optimal matrix from the most recent iteration as a starting point. Finally I obtained a better design, with $\det[\boldsymbol{M}(\xi,\boldsymbol{\beta})] = 3.368853 \times 10^{-7}$. The failure of the value of the determinant to steadily increase with recent iterations of the program made me wonder whether perhaps I was close to the locally D-optimal design, so I decided to investigate the value of the standardised variance at and near the support points of the current design using Program_13 (see page 113). However, the output from this program included the following:

```
For support point  10
 Maximum std var is  9.465257  at
 1 1 0.8605391 0.6148241 -0.9906819 -0.6833456
```

This was disheartening, as one would want the maximum to be essentially $p = 9$.

I could have continued searching, perhaps after changing to -1 or 1 those coordinates of each support point that were essentially equal to (say) -0.999 or 0.999. However, in the interests of demonstrating an alternative approach to locating a locally D-optimal design, I instead decided to start again, this time from the minimum value of s: nine.

With this amendment, I used Segments ①,② and ③ of the program beginning on page 106. After several iterations, I replaced the last line of Segment ⑤ immediately above by `wtout <- rep(1/9,9)` as I knew that, if $s = 9$ is correct, the values of $\delta_1, \ldots, \delta_9$ must all be 1/9 (see page 71). Keeping this amendment, I ran the program several times, each time using the current "best" design as the basis for the input for the next search. However, I could not find a design with a value of $\det[\boldsymbol{M}(\xi, \boldsymbol{\beta})]$ greater than 1.728427×10^{-7}. As this is less than the maximum value of 3.368853×10^{-7} found for $s = 12$, I decided to give up on a search for $s = 9$.

Six of the support points for the best design for $s = 9$ had each coordinate equal to -1 or 1. As my experience is that such points are often part of a locally D-optimal design, I decided to incorporate these six points in the initial guess of a design for $s = 10$. I generated the other four points randomly, and also generated the design weights randomly. I then followed a similar procedure to that described above, namely using as an initial search point for *optim* slight variations around the best design obtained in 2000 simulations from the previous design. In 18 repeats of this procedure, the greatest value of $\det[\boldsymbol{M}(\xi, \boldsymbol{\beta})]$ rose steadily from 8.628362×10^{-11} to 2.159584×10^{-7}, but would not improve further. Again, this was clearly a worse situation than for $s = 12$, so I decided to try $s = 11$ support points.

For initial searches, I generated the values of *variable* (to be transformed to 11 6-dimensional support points with each coordinate satisfying $-1 \leq x_{ij} \leq 1$) randomly. However, I set the design weights equal to 1/11 for each support point, in an attempt to prevent any support point from "disappearing" on the first search. Two thousand iterations gave me a design with $\det[\boldsymbol{M}(\xi, \boldsymbol{\beta})] = 1.110171 \times 10^{-9}$. This value was greater than the *initial* values for searches with other values of s, so I decided to use variations of this design as starting points as input to *optim*. Twenty repetitions of this procedure led me to $\det[\boldsymbol{M}(\xi, \boldsymbol{\beta})] = 3.460214 \times 10^{-7}$, but then I could not obtain any greater values of $\det[\boldsymbol{M}(\xi, \boldsymbol{\beta})]$. In one sense, this was not too worrying, because I had now obtained a value greater than the best value obtained for $s = 12$. The design weight for the 10th support point of this design was 1.4667×10^{-5}, suggesting that this point could be removed. However, simply removing it in the manner described previously for transitioning from 15 to 13 support points did not lead to an improved design. I then noticed that the remaining 10 support points from this design had every coordinate virtually equal to -1 or 1. So I decided to round each coordinate to -1 or 1, and to use these 10 points as fixed points in a design for $s = 10$ support points, and to let only the design weights vary when using *optim*.

Recall from Segment ③ on page 123 that *out4* is the $s \times (m+1)$ matrix that forms the output from *optim* after conversion back to constrained data. Each row represents a support point, and the first m elements are the values of the m explanatory variables at that point. The $(m + 1)$th element is the value of the design weight.

I created the $s \times m$ matrix of coordinates *xmat* by rounding each coordinate from the previous design to -1 or 1. This matrix will not be changed by *optim*. Then I modified the function *detinfomat* to produce *detinfomat3*, which accepts as input a vector of s z-values that will be converted to design weights. The function *optim* will search for the particular set of z-values (and hence for the design weights) for which $\det[\boldsymbol{M}(\xi, \boldsymbol{\beta})]$ is maximised.

```
out4 <- spare
xmat <- round(out4[c(1:9,11),1:6],1)

detinfomat3 <- function(variables)
{
 wtvals <- (variables)^2
```

```
 deltavec <- wtvals/sum(wtvals)
 fxmat <- apply(xmat,2,fx)
 etavec <- as.vector(t(betavec)%*%fxmat)
 expetavec <- exp(etavec)
 modelwtvec <- expetavec/((1+expetavec)^2)
 infomat <- fxmat %*%diag(deltavec*modelwtvec)%*%t(fxmat)
 -det(infomat)
}

nsims <- 2000
mindet <- 100

for (i in 1:nsims)
{
 initial <- runif(s)
 out <- optim(initial,detinfomat3,NULL,method="Nelder-Mead")
 valuenow <- out$val
 if(valuenow < mindet){mindet <- valuenow
 design <- out$par}
}
cat("Min value of det\n",mindet,"\n")
output <- design
out1 <- cos(pi*initialz1)
out2 <- (output)^2
wts <- out2/sum(out2)
out4 <- cbind(matrix(out1,s,m),wts)
cat("Design\n")
t(out4)
```

The result was gratifying. The maximum value of $\det[\boldsymbol{M}(\xi, \boldsymbol{\beta})]$ increased to 3.534999×10^{-7}. I did try another search, using minor variations of the design weights as the input to *optim*, but the change in $\det[\boldsymbol{M}(\xi, \boldsymbol{\beta})]$ was negligible, and the design weights did not change at all when written to three decimal places. So I elected to stay with the first of the two designs.

I then ran Program_13 (on page 113) to calculate the standardised variance at each support point of this design, and to investigate the maximum value of the standardised variance in the neighbourhood of each support point. The output was as follows:

```
at support point  1  stdvar =  8.999653
at support point  2  stdvar =  9.001946
at support point  3  stdvar =  9.001332
at support point  4  stdvar =  8.994732
at support point  5  stdvar =  8.997475
at support point  6  stdvar =  8.999568
at support point  7  stdvar =  9.00329
at support point  8  stdvar =  8.998748
at support point  9  stdvar =  8.999784
at support point  10  stdvar =  9.004897
```

```
For support point  1
 Maximum std var is  9.002609  at
 -1 -1 1 -1 -1 0.9801071
For support point  2
 Maximum std var is  9.001946  at
 1 -1 -1 -1 1 1
For support point  3
 Maximum std var is  9.001332  at
 1 1 1 -1 1 -1
For support point  4
 Maximum std var is  8.994732  at
 1 -1 -1 1 -1 -1
For support point  5
 Maximum std var is  8.997475  at
 -1 -1 -1 1 1 1
For support point  6
 Maximum std var is  8.999568  at
 1 1 1 1 -1 1
For support point  7
 Maximum std var is  9.00329  at
 -1 -1 -1 -1 -1 -1
For support point  8
 Maximum std var is  8.998748  at
 -1 1 1 1 1 -1
For support point  9
 Maximum std var is  8.999784  at
 -1 1 1 -1 -1 1
For support point  10
 Maximum std var is  9.004897  at
 1 -1 1 1 1 1
```

The values of the standardised variances at each support point are all effectively equal to $p = 9$, and nowhere in the neighbourhood of any support point is there a potential support point with a standardised variance greater than 9. By the general equivalence theorem, I conclude that this is a locally D-optimal design. The actual design is given in Table 4.4.

In this section, I have detailed all of the tedium involved in obtaining the locally D-optimal design, rather than simply present the answer. You need to see this at least once, so that you know what to expect. While I started my search by checking carefully the minimum value of the design weights for each design given by 2000 calls to *optim*, by the end I had become mesmerised by the value of $\det[\boldsymbol{M}(\xi, \boldsymbol{\beta})]$ after each set of searches. This was a mistake. If I had continued to look at the minimum value of $\delta_1, \ldots, \delta_s$, I would have recognised much earlier in my search with $s = 11$ that, in fact, one support point could be dropped and the remaining support points all essentially had only -1 or 1 as their coordinates. This would have saved time.

I consider that it was just bad luck that my searches for the optimal design with $s = 10$ failed to find that design. With hindsight, I became stuck at a local maximum rather than the overall maximum. How might I have avoided this? Perhaps I could

Point	x_1	x_2	x_3	x_4	x_5	x_6	δ
1	-1	-1	1	-1	-1	1	0.087
2	1	-1	-1	-1	1	1	0.093
3	1	1	1	-1	1	-1	0.109
4	1	-1	-1	1	-1	-1	0.110
5	-1	-1	-1	1	1	1	0.109
6	1	1	1	1	-1	1	0.110
7	-1	-1	-1	-1	-1	-1	0.099
8	-1	1	1	1	1	-1	0.097
9	-1	1	1	-1	-1	1	0.099
10	1	-1	1	1	1	1	0.087

Table 4.4 *Locally D-optimal design when there are $m = 6$ explanatory variables, there is a logit link, and the linear predictor is $\eta = 0.4 + 0.3x_1 + 0.4x_2 - 0.5x_3 - 0.2x_4 + 0.3x_5 + 0.4x_6 + 0.2x_1x_2 - 0.3x_3x_4$.*

have generated still more random initial inputs to *optim*, and then compensated for this by running less variations on the "best so far" solution when initiating subsequent searches. Or perhaps, when I became stuck with $s = 12$, I should have gone to $s = 11$ instead of choosing to go to $s = 9$ in order to give a better demonstration of various techniques. You be the judge. Just remember that, when seeking to optimise a complicated function of many variables where the solution is likely to incorporate values on the boundary of the design space, obtaining the optimal solution is uncertain, and very likely to be a slow process.

4.7 Obtaining an exact design

All the D-optimal designs found so far in this chapter have been approximate designs; see page 50. However, practitioners will want exact designs, where the design weights are a multiple of $1/N$, where N is the total number of observations to be taken. The first way that this may be done is to convert the design weights $\delta_1, \ldots, \delta_s$ of the approximate design to numbers of observations $n_1, \ldots, n_s$ (with $\sum_i n_i = N$) using the method of Pukelsheim (1993, Chapter 12), as described and illustrated in Sub-section 3.3.2. Program_6 in the online resources will do the computations for you.

For $m = 2$, consider the logit link and the linear predictor $\eta = 2 + 4x_1 + 3x_2$. From the online tables, the locally D-optimal (approximate) design is

x_1	x_2	δ
−1.000	0.213	0.149
−1.000	1.000	0.306
−0.092	−1.000	0.236
0.594	−1.000	0.309

.

Suppose that it is proposed to run the design with $N = 6$ observations. The first two commands of Program_6 are

```
N <- 6
deswts <- c(0.149,0.306,0.236,0.309)
```

The unique apportionment given by the program is

```
[1] 1 2 1 2
```

meaning that the four support points above are given weights 1/6, 2/6, 1/6 and 2/6, respectively.

Another approach is to select the value of s and to give a design weight of $1/s$ to each point, which will have the effect of making N a multiple of s. One then modifies the function *detinfomat2* so that the design weights are fixed at $1/s$ each and the ms elements of the vector that is given as input to *detinfomat* provide the elements of the support points $\mathbf{x_1}, \ldots, \mathbf{x_s}$. The procedure was described in Section 3.3.3, and is now illustrated for $s = 6$.

The following program was run:

```
betavec <- c(2,4,3)
p <- length(betavec)
m <- 2
s <- 6
lim1 <- m*s
deltavec <- rep(1/s,s)

fx <- function(xvec)
{
 fvec <- c(1,xvec)
 fvec
}

detinfomat3 <- function(variables)
{
 xmat <- matrix(cos(pi*variables),m,s,byrow=T)
 fxmat <- apply(xmat,2,fx)
 etavec <- as.vector(t(betavec)%*%fxmat)
 expetavec <- exp(etavec)
 modelwtvec <- expetavec/((1+expetavec)^2)
 infomat <- fxmat%*%diag(deltavec*modelwtvec)%*%t(fxmat)
 -det(infomat)
}

#simulations of different initial values
nsims <- 2000
mindet <- 10
for (i in 1:nsims)
{
 initial <- runif(lim1)
 out <- optim(initial,detinfomat3,NULL,method="Nelder-Mead")
 valuenow <- out$val
 if(valuenow < mindet){mindet <- valuenow
 design <- out$par}
}
cat("Min value of det\n",mindet,"\n")
```

```
output <- design
out1 <- cos(pi*output[1:lim1])
out4 <- cbind(matrix(out1,s,2),deltavec)
out4 <- out4[order(out4[,1],out4[,2]),]
cat("Design\n")
out4
t(out4)
```

The function *detinfomat3* is very similar to *detinfomat2* (see page 107), except that its input consists of a vector of only ms elements, rather than $m(s+1)$ elements, and it does not calculate the vector, *deltavec*, of design weights, because these are fixed and are defined externally to the program. The output was similar to the following:

```
Min value of det
 -0.0003814488

Design
> t(out4)
              [,1]      [,2]      [,3]       [,4]      [,5]      [,6]
         -0.999997 -0.999701 -0.998625 -0.0953040  0.554944  0.565829
          0.999037  0.999012  0.161108 -0.9999946 -0.999681 -0.999844
deltavec  0.166667  0.166667  0.166667  0.1666667  0.166667  0.166667
```

Note that support points $\boldsymbol{x}_1$ and $\boldsymbol{x}_2$ are essentially the same, and $\boldsymbol{x}_5$ and $\boldsymbol{x}_6$ are very similar. This suggests to me that probably points 1 and 2 will merge to one point with a weight of 2/6, and that support points 5 and 6 will merge to a second point, also with a weight of 2/6. However, I decided to use deviations from the current solution (stored as *output*), with the same value of *deltavec* as before, to search for an improved design. The next search began as follows:

```
nsims <- 2000
mindet <- 10
initialz1 <- output
for (i in 1:nsims)
{
 initial <- initialz1 + 0.1*(runif(lim1)-0.5)
 out <- optim(initial,detinfomat3,NULL,method="Nelder-Mead")
 valuenow <- out$val
 if(valuenow < mindet){mindet <- valuenow
 design <- out$par}
}
```

Some of the output appears below:

```
Min value of det
 -0.0003819232

> t(out4)
              [,1]      [,2]      [,3]      [,4]      [,5]      [,6]
         -1.000000 -1.000000 -0.999998 -0.102959  0.564502  0.564629
          1.000000  1.000000  0.156122 -1.000000 -1.000000 -1.000000
deltavec  0.166667  0.166667  0.166667  0.166667  0.166667  0.166667
```

This reinforces the conclusion drawn before, that probably $\boldsymbol{x}_1 = \boldsymbol{x}_2$ and $\boldsymbol{x}_5 = \boldsymbol{x}_6$. So I did one more search, specifying as an initial estimate a random variation on the four distinct support points above, and giving a revised value of *deltavec*. Remember that this requires the values of s and ms to be altered! The program begins as follows:

```
s <- 4
lim1 <- m*s

initialx <- c(-1,-1,-0.103,0.565,1,0.156,-1,-1)
initialz1 <- acos(initialx)/pi
deltavec <- c(2,1,1,2)/6
nsims <- 2000
mindet <- 10
for (i in 1:nsims)
{
 initial <- initialz1 + 0.1*(runif(lim1)-0.5)
 out <- optim(initial,detinfomat3,NULL,method="Nelder-Mead")
 valuenow <- out$val
 if(valuenow < mindet){mindet <- valuenow
 design <- out$par}
}
```

Segments of the output appear below:

```
 -0.0003819235
> t(out4)
               [,1]       [,2]       [,3]       [,4]
         -1.0000000 -1.0000000 -0.1027178  0.5645266
          1.0000000  0.1558978 -1.0000000 -1.0000000
deltavec  0.3333333  0.1666667  0.1666667  0.3333333
```

So we have two potential exact designs from this sub-section, one using Pukelsheim's method and one fixing the weights and then searching for the best support points. Let us call them Designs A (ξ_A) and B (ξ_B), respectively. They appear below.

Design A			*Design B*		
x_1	x_2	δ	x_1	x_2	δ
-1.000	0.213	1/6	-1.000	0.156	1/6
-1.000	1.000	2/6	-1.000	1.000	2/6
-0.092	-1.000	1/6	-0.103	-1.000	1/6
0.594	-1.000	2/6	0.565	-1.000	2/6

The two designs are not dissimilar. Neither is locally D-optimal, because neither has been selected through optimising $\det[\boldsymbol{M}(\xi, \boldsymbol{\beta})]$ with regard to the support points and the design weights collectively. Which is the better? From (3.17),

$$\text{D-efficiency of } \xi_A \text{ relative to } \xi_B = \left\{ \frac{\det[M(\xi_A, \boldsymbol{\beta})]}{\det[M(\xi_B, \boldsymbol{\beta})]} \right\}^{1/p}.$$

It is straightforward to use *detinfomat2* (page 107) to calculate that $\det[M(\xi_A, \boldsymbol{\beta})] = 3.128805 \times 10^{-5}$ and $\det[M(\xi_B, \boldsymbol{\beta})] = 3.872962 \times 10^{-5}$, and so the

$$\text{D-efficiency of } \xi_A \text{ relative to } \xi_B = \left[\frac{3.128805 \times 10^{-5}}{3.872962 \times 10^{-5}} \right]^{1/3} = 0.931.$$

Alternatively, we may say that the D-efficiency of ξ_B relative to ξ_A is 1.074.

Consequently, of the two exact designs obtained here, Design B would be preferred.

4.8 When sample sizes are small

4.8.1 Maximum penalised likelihood estimation

Recall from Section 1.5 that the model parameters $\beta_0, \ldots, \beta_{p-1}$ are generally estimated by the ML estimators, which maximise the likelihood given in (1.16). For very large values of N, the vector of estimators, $\hat{\boldsymbol{\beta}}$, satisfies $\mathrm{E}(\hat{\boldsymbol{\beta}}) = \boldsymbol{\beta}$ and $\mathrm{cov}(\hat{\boldsymbol{\beta}}) = \boldsymbol{\mathcal{I}}^{-1}$.

If N is "small," these results may be inaccurate. In particular, the result for $\mathrm{cov}(\hat{\boldsymbol{\beta}})$, on which the calculation of a locally D-optimal design is based, may lead to a design which is not as desirable as hoped. To reduce the bias of each $\hat{\beta}_i$ (discrepancy between $\mathrm{E}(\hat{\beta}_i)$ and the true value β_i), $i = 0, \ldots, p-1$, Firth (1993) introduced the *maximum penalised likelihood* (MPL), $L^*(\boldsymbol{\beta}; y_1, \ldots, y_N)$, whose logarithm (denoted by ℓ^*) is given by

$$\ell^*(\boldsymbol{\beta}; y_1, \ldots, y_N) = \ln L(\boldsymbol{\beta}; y_1, \ldots, y_N) + \frac{1}{2}\det\left[\mathcal{I}(\boldsymbol{\beta}; y_1, \ldots, y_N)\right]$$
$$= \ell(\boldsymbol{\beta}; y_1, \ldots, y_N) + \frac{1}{2}\det\left[\mathcal{I}(\boldsymbol{\beta}; y_1, \ldots, y_N)\right]. \tag{4.21}$$

Example 4.8.1. *Consider a Bernoulli random variable, Y, with $P(Y_i = 1) = \pi_i$ $(i = 1, 2)$. Model π_i using $\ln[\pi_i/(1-\pi_i)] = \eta_i = \beta_0 + \beta_1 x_i$. Take n_i observations at $x = x_i$, and write y_i for the total of those observations $(i = 1, 2)$. This scenario was investigated in detail by Russell et al. (2009a).*

The MPL estimators of β_0 and β_1 are the values, β_0^* and β_1^*, that maximise ℓ^*. I will occasionally write them as $\beta_0^*(y_1, y_2)$ and $\beta_1^*(y_1, y_2)$ to show their dependence on y_1 and y_2. They can be shown to be

$$\beta_1^*(y_1, y_2) = \frac{\ln[(y_1+0.5)/(n_1-y_1+0.5)] - \ln[(y_2+0.5)/(n_2-y_2+0.5)]}{x_1 - x_2} \tag{4.22}$$

$$\beta_0^*(y_1, y_2) = \frac{x_1 \ln[(y_2+0.5)/(n_2-y_2+0.5)] - x_2 \ln[(y_1+0.5)/(n_1-y_1+0.5)]}{x_1 - x_2}. \tag{4.23}$$

Compare these estimates with the ML estimates given in (3.14) *and* (3.15). *The only difference is the presence of the "+0.5" in various places in the MPL estimates. For small values of n_1 and n_2, they will induce a difference between the ML and MPL estimates, but as n_1 and n_2 get larger, their effect diminishes. Asymptotically, the MPL and ML estimates will be equivalent. So the matrix $\mathcal{I}^{-1}$ serves as the asymptotic covariance matrix of the MPL estimates as well as the ML estimates, and $\mathbf{M}(\xi, \boldsymbol{\beta})$ can be used to determine the locally D-optimal design for small n_1 and n_2, provided that one is prepared to risk the occurrence of errors introduced by using an asymptotic result when it is not appropriate.*

Note from (4.22) and (4.23) that β_0^* and β_1^* do not become infinite when y_i equals 0 or n_i (i.e., when $\bar{y}_i$ equals 0 or 1). So the problem of separation (see Section 3.6) does not prevent us from obtaining finite estimates of β_0 and β_1.

Just as was noted on page 65 when considering $\hat{\beta}_0$ and $\hat{\beta}_1$, it is not possible to find explicit expressions for β_0^* and β_1^* when $s > 2$. An iterative solution is required. The `glm` function in R cannot be used because it does not cope with the separation problem, but you can use `brglm`. It is necessary to install the package and then use the command `library(brglm)` before you can use the function, but this is simple to do. Then use `brglm` in a similar manner to `glm`. However, note that it works only for the *binomial* distribution. If your work relates only to data from Bernoulli or binomial distributions, then I suggest that you use `brglm`. It is easy to download, install, and run. However, if you may need to perform analyses for data from Poisson, multinomial or other distributions, the package `brglm2` will be more useful. It is described in Section 5.4.

Example 4.8.2. *Page 66 contained the result of an analysis by* `glm` *of an experiment where* $n_1 = n_2 = 5$, $x_1 = 0$, $x_2 = 1$, $\bar{y}_1 = 0/5$ *and* $\bar{y}_2 = 3/5$. *The output suggested that a separation exists in the data. Using the commands*

```
library(brglm)
x <- c(0,1)
nvec <- c(5,5)
yvec <- c(0,3)
fail <- nvec - yvec
model <- brglm(cbind(yvec,fail)~x,family=binomial("logit"))
summary.brglm(model)$coefficients
```

results in the output

```
              Estimate Std. Error   z value  Pr(>|z|)
(Intercept) -2.397895   1.618080 -1.481939 0.1383565
x            2.734368   1.855004  1.474050 0.1404682
```

This is very different from the output on page 66.

Example 4.8.3. *Compare the output from* `glm` *and* `brglm` *for an experiment with* $s = 3$ *support points,* $x_1 = 0$, $x_2 = 0.5$, $x_3 = 1$, $n_1 = n_2 = n_3 = 4$ *and* $y_{1\cdot} = 0$, $y_{2\cdot} = 4$ *and* $y_{3\cdot} = 1$. *With* `glm`, *one gets*

```
             Estimate Std. Error z value Pr(>|z|)
(Intercept)  -0.8783     0.9881  -0.889    0.374
x             1.0529     1.4856   0.709    0.479
```

while, with `brglm`, *one obtains*

```
             Estimate Std. Error z value Pr(>|z|)
(Intercept)  -0.7277     0.9646  -0.754    0.451
x             0.8604     1.4628   0.588    0.556
```

Example 4.8.4. *Again consider an experiment with* $s = 3$ *support points,* $x_1 = 0$, $x_2 = 0.5$, $x_3 = 1$, $n_1 = n_2 = n_3 = 4$ *and now take* $y_{1\cdot} = 1$, $y_{2\cdot} = 2$ *and* $y_{3\cdot} = 3$. *With* `glm`, *one obtains*

```
             Estimate Std. Error   z value  Pr(>|z|)
(Intercept) -1.098612   1.032796 -1.063727 0.2874525
x            2.197225   1.632993  1.345520 0.1784574
```

while, with `brglm`, *one gets*

```
              Estimate Std. Error    z value  Pr(>|z|)
(Intercept) -0.8848655   0.990532 -0.8933235 0.3716839
x            1.7697311   1.554900  1.1381639 0.2550521
```

Each of Examples 4.8.2 to 4.8.4 illustrates that there are nontrivial differences between the values of the ML estimates and corresponding MPL estimates, and that the estimated standard errors of the MPL estimates are less than those of the ML estimates. This latter relationship is not a coincidence, but is a property of the MPL procedure.

Example 4.8.5. *Suppose that we use the same three support points as in Example 4.8.4, but take* $n_i = 40$ *observations at each point and obtain* $y_{1\cdot} = 10$, $y_{2\cdot} = 20$ *and* $y_{3\cdot} = 30$. *That is, each sample size has been increased by a factor of 10, but the proportions of successes at each support point are exactly as before. The results from the two analyses are as follows:*

glm

```
             Estimate Std. Error   z value     Pr(>|z|)
(Intercept) -1.098612  0.3265986 -3.363799 7.687739e-04
x            2.197225  0.5163978  4.254907 2.091358e-05
```

brglm

```
             Estimate Std. Error   z value     Pr(>|z|)
(Intercept) -1.072588  0.3248144 -3.302157 9.594443e-04
x            2.145176  0.5130817  4.180964 2.902761e-05
```

As might be expected from the proportions of successes at each support point being unchanged, the ML estimates are unchanged from Example 4.8.4. However, the increase in replication by a factor of 10 has caused their estimated standard errors to be decreased by a factor of $\sqrt{10}$. The increases in sample size have reduced the bias, and so the MPL estimates are now closer to the ML estimates than they were for smaller sample sizes. The estimated standard errors of the ML estimators are now quite similar to those of the corresponding MPL estimators.

Although the MPL estimates of β_0 and β_1 have been calculated explicitly for the logit link only, separation also occurs with the probit and complementary log-log distributions. The output below shows the results obtained for the same experimental results considered on page 66, in the experiment where $n_1 = n_2 = 5$, $x_1 = 0$, $x_2 = 1$, $\bar{y}_{1\cdot} = 0/5$ and $\bar{y}_{2\cdot} = 3/5$. The ML estimates, and then the MPL estimates, for the probit link are shown first, followed by analogous results for the complementary log-log link. In each case, one sees very large values of the estimated standard errors for the ML estimates, and much smaller values for the estimated standard errors of the MLP estimates.

Probit link

```
> model <- glm(cbind(yvec,fail)~x,family=binomial("probit"))
> summary.glm(model)$coefficients
             Estimate Std. Error       z value  Pr(>|z|)
(Intercept) -6.533657   7290.271 -0.0008962159 0.9992849
x            6.787004   7290.271  0.0009309673 0.9992572
> model2 <- brglm(cbind(yvec,fail)~x,family=binomial("probit"))
> summary.brglm(model2)$coefficients
             Estimate Std. Error   z value   Pr(>|z|)
(Intercept) -1.466229  0.8450781 -1.735022 0.08273682
x            1.684543  1.0167652  1.656767 0.09756654
```

Complementary log-log link

```
> model <- glm(cbind(yvec,fail)~x,family=binomial("cloglog"))
> summary.glm(model)$coefficients
             Estimate Std. Error       z value  Pr(>|z|)
(Intercept) -24.51075    56991.3 -0.0004300787 0.9996568
x            24.42332    56991.3  0.0004285448 0.9996581
> model2 <- brglm(cbind(yvec,fail)~x,family=binomial("cloglog"))
> summary.brglm(model2)$coefficients
             Estimate Std. Error   z value  Pr(>|z|)
(Intercept) -2.397895   1.517597 -1.580061 0.1140930
x            2.323253   1.630424  1.424938 0.1541752
```

When small values of N are to be used, how do we choose an optimal design? We could use conventional D-optimality, but there are risks associated with assuming that $\mathrm{E}(\hat{\boldsymbol{\beta}}) = \boldsymbol{\beta}$ and that $\mathrm{cov}(\hat{\boldsymbol{\beta}}) = \boldsymbol{M}^{-1}(\xi, \boldsymbol{\beta})$, particularly given that separation may be likely to occur.

Russell et al. (2009a) suggested an alternative approach for small N, where the use of the MPL estimator $\boldsymbol{\beta}^*$ is indicated. They argued that, as the parameters in $\boldsymbol{\beta}$ are frequently estimated in order to predict the probability of a "success" on a single observation at x, $\pi(x) = 1/\{1 + \exp[-\eta(x)]\}$, attention should be concentrated on properties of the MPL estimator of this probability:

$$\pi^*(x) = 1/\{1 + \exp[-\boldsymbol{f}^\top(x)\boldsymbol{\beta}^*]\}.$$

We consider the *mean square error* (MSE) of $\pi^*(x)$. It is defined by

$$\mathrm{MSE}[\pi^*(x)] = \mathrm{E}\left\{[\pi^*(x) - \pi(x)]^2\right\}, \tag{4.24}$$

the expected value of the square of the difference between the estimator of $\pi(x)$ and the actual value. We would like to minimise this. The MSE is an appropriate measure to use when N is small and the MPL estimator $\boldsymbol{\beta}^*$ may be biased. It is a standard result in most introductory texts in mathematical statistics that

$$\mathrm{MSE}[\pi^*(x)] = [\text{bias of } \pi^*(x)]^2 + \mathrm{Var}[\pi^*(x)].$$

Hopefully, minimising the MSE of $\pi^*(x)$ will give a design with the property that the estimator has small bias and small variance; choosing a design that minimises the variance alone may result in an estimator with large bias being produced.

Let us consider how this involves the selection of a design. For a given parameter vector $\boldsymbol{\beta} = (\beta_0, \beta_1)^\top$, choose two support points x_1 and x_2, and take n_1 observations at x_1 and n_2 observations at x_2, where $n_1 + n_2 = N$. At x_i, the probability of a "success" on an individual observation is $\pi_i = 1/\{1 + \exp[-\boldsymbol{f}^\top(x_i)\boldsymbol{\beta}]\}$. If Y_i is the number of successes observed at x_i, then $Y_i \sim \mathrm{Bin}(n_i, \pi_i)$ $(i = 1, 2)$. The observed values (y_1, y_2) lead to MPL estimates β_0^* and β_1^* by Equations (4.22) and (4.23). Then the probability of a success at a point x is

$$\pi(x) = 1/\{1 + \exp[-\eta(x)]\} = 1/[1 + \exp(-\beta_0 - \beta_1 x)],$$

and the estimate obtained by using the MPL estimators is

$$\pi^*(x, y_1, y_2) = 1/\{1 + \exp[-\beta_0^*(y_1, y_2) - \beta_1^*(y_1, y_2)x]\}.$$

(I have written this probability as $\pi^*(x, y_1, y_2)$ because it depends on the values of y_1 and y_2 that are used to calculate β_0^* and β_1^*.)

The probability of obtaining this particular value of $\pi^*(x, y_1, y_2)$ is the probability that the particular pair (y_1, y_2) was observed, namely

$$\begin{aligned} \mathrm{P}(Y_1 = y_1, Y_2 = y_2) &= \mathrm{P}(Y_1 = y_1) \times \mathrm{P}(Y_2 = y_2) \\ &= \binom{n_1}{y_1}\pi_1^{y_1}(1 - \pi_1)^{n_1 - y_1} \times \binom{n_2}{y_2}\pi_2^{y_2}(1 - \pi_2)^{n_2 - y_2}. \end{aligned}$$

The value of $\mathrm{E}\left\{[\pi^*(x) - \pi(x)]^2\right\}$ is calculated by finding $[\pi^*(x, y_1, y_2) - \pi(x)]^2$ for a given value of (y_1, y_2), multiplying this by $\mathrm{P}(Y_1 = y_1, Y_2 = y_2)$, and summing these products over all possible values of (y_1, y_2). That is,

$$\begin{aligned} \mathrm{MSE}[\pi^*(x)] = &\sum_{y_1=0}^{n_1}\sum_{y_2=0}^{n_2}[\pi^*(x, y_1, y_2) - \pi(x)]^2 \\ &\times \binom{n_1}{y_1}\pi_1^{y_1}(1 - \pi_1)^{n_1 - y_1} \times \binom{n_2}{y_2}\pi_2^{y_2}(1 - \pi_2)^{n_2 - y_2}. \end{aligned} \tag{4.25}$$

For a given value of $\boldsymbol{\beta}$ and a design which specifies the values of n_1, n_2, x_1 and x_2, $\text{MSE}[\pi^*(x)]$ is a function of x alone. For simplicity, it will be written as $\text{MSE}(x)$.

Over what values of x should $\text{MSE}(x)$ be considered? It is tempting to say "from $-\infty$ to ∞", but this is not feasible, as $\pi^*(x)$ and $\pi(x)$ would be expected to become very small as x approaches these extremities. The approach taken by Russell et al. (2009a) was to choose values $x_{0.0001}$ and $x_{0.9999}$ satisfying $\pi(x_{0.0001}) = 0.0001$ and $\pi(x_{0.9999}) = 0.9999$ on the grounds that these cover almost all useful values of x. Figure 4.8 illustrates this.

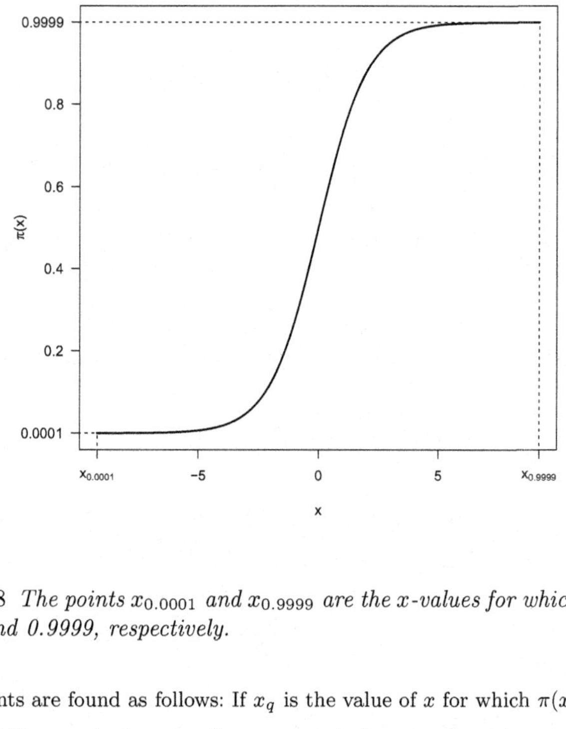

Figure 4.8 *The points $x_{0.0001}$ and $x_{0.9999}$ are the x-values for which $\pi(x)$ equals 0.0001 and 0.9999, respectively.*

These points are found as follows: If x_q is the value of x for which $\pi(x) = q$, then

$$\pi(x_q) = 1/[1+\exp(-\beta_0 - \beta_1 x_q)] = q \Rightarrow \exp(-\beta_0 - \beta_1 x_q) = 1/q - 1$$

$$\Rightarrow -\beta_0 - \beta_1 x_q = \ln(1/q - 1) = -\ln\left[\frac{q}{1-q}\right]$$

$$\Rightarrow x_q = \{\ln\left[\frac{q}{1-q}\right] - \beta_0\}/\beta_1.$$

For the case of $\beta_0 = 0$ and $\beta_1 = 1$ (which is equivalent to the canonical variable $\eta = \beta_0 + \beta_1 x$), it follows that $\eta_{0.0001} = -9.21024$ and $\eta_{0.9999} = 9.21024$.

A consideration of a function to calculate $\text{MSE}(x)$ will be deferred until later. Consider first the appearance of a plot of $\text{MSE}(x)$ vs x over $\{x\colon -9.21024 \leq x \leq 9.21024\}$. Take $\boldsymbol{\beta} = (0,1)^{\mathsf{T}}$, and let the support points be $x_1 = -1.5434$ and $x_2 = 1.5434$, which are the support points of $\xi^*_{\text{L},\,z}$ in (4.8).

Figure 4.9 compares graphs of MSE(x) for $n_1 = n_2 = 3$, 6, 9 and 18. A graph for $n_1 = n_2 = 4$, 5, 10 and 20 can be seen in Russell et al. (2009a, Figure 2). It is clear that the number of observations at each support point influences the value of the MSE considerably.

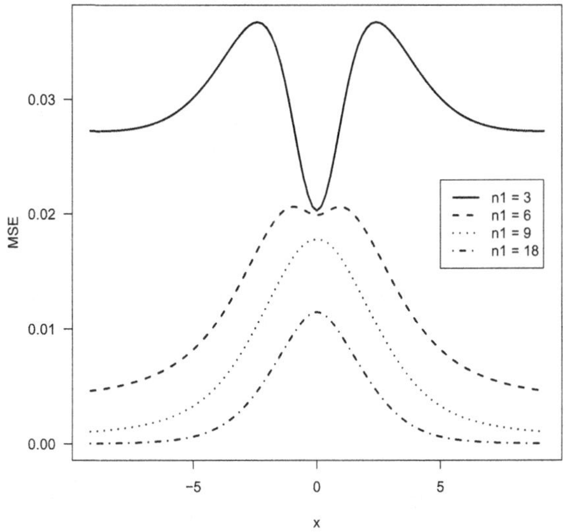

Figure 4.9 *For* $\boldsymbol{\beta} = (0, 1)^\top$ *and* $x_1 = -1.5434$, $x_2 = 1.5434$, *plots of MSE*(x) *vs.* x *for* $n_1 = n_2$ *with* $n_1 = 3, 6, 9$ *and* 18.

Consideration should also be given to the influence of the support points on the value of MSE(x). For $\boldsymbol{\beta} = (0, 1)^\top$ and $n_1 = n_2 = 6$, Figure 4.10 repeats the plot of MSE(x) vs. x for the support points $x_1 = -1.5434$ and $x_2 = 1.5434$, and also shows a plot of MSE(x) vs. x for the support points $x_1 = -1.0000$ and $x_2 = 2.0000$. The two plots are very different.

4.8.2 IMSE-optimality

To compare designs on the basis of plots of MSE(x) vs x, it is usual to compare the areas that are beneath the curves and above the horizontal axis. That is, for each plot we consider

$$\int_{-\infty}^{\infty} \mathrm{MSE}(x)\, dx \approx \int_{x_{0.0001}}^{x_{0.9999}} \mathrm{MSE}(x)\, dx, \tag{4.26}$$

which is known as the *integrated mean square error* (IMSE) . From amongst a set of candidate designs, the design for which the IMSE is least is said to be *IMSE-optimal.*

As it is not possible to write an expression for MSE(x) in terms of x, it becomes necessary to evaluate the integral in (4.26) using numerical integration. As Figures 4.9 and 4.10 suggest that the curve MSE(x) is reasonably smooth, Simpson's rule (see Section 2.5) will be used to evaluate (4.26).

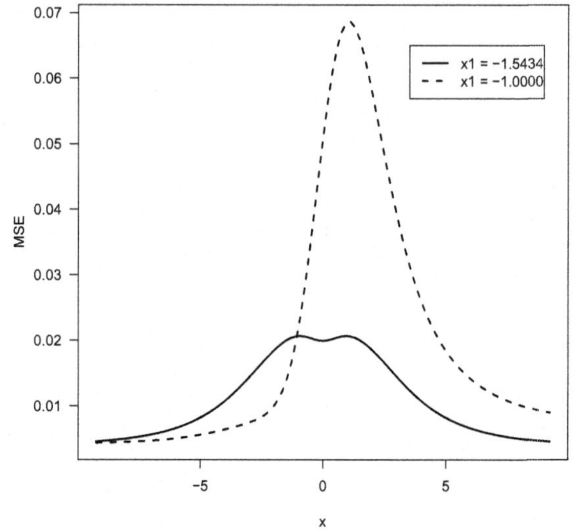

Figure 4.10 *For* $\boldsymbol{\beta} = (0, 1)^\top$ *and* $n_1 = n_2 = 6$*, plots of MSE*(x) *vs.* x *for two sets of support points:* $(x_1, x_2) = (-1.5434, 1.5434)$ *and* $(-1.0000, 2.0000)$.

Example 4.8.6. *Consider the case of* $n_1 = n_2 = 6$, $x_1 = -1.5434$ *and* $x_2 = 1.5434$. *The following program, which also appears within Program_14 in the Web site doeforglm.com, will calculate the values of MSE*(x) *at 101 equally spaced points between* -9.21024 *and* 9.21024.

①

```
fx <- function(x)
{
 c(1,x)
}

mse <- function(x,x1,x2)
{#Calculates the MSE at a value x for support points x1 and x2
 distx <- x1 - x2
 eta1 <- sum(fx(x1)*betavec)
 eta2 <- sum(fx(x2)*betavec)
 prob1 <- 1/(1+exp(-eta1))
 prob2 <- 1/(1+exp(-eta2))
 y1probs <- dbinom(y1vals,n1,prob1)
 y2probs <- dbinom(y2vals,n2,prob2)
```

```
 pix <- 1/(1+exp(-sum(fx(x)*betavec)))
```

②

```
 sumi <- 0
 for (i in 1:(n1+1))
 {
  sumj <- 0
  for (j in 1:(n2+1))
   {
    beta0star <- (a1star[i]*x1-a2star[j]*x2)/distx
    beta1star <- (a2star[j] - a1star[i])/distx
    pistarx <- 1/(1+exp(-beta0star - beta1star*x))
    sumj <- sumj + ((pistarx-pix)^2)*y2probs[j]
   }
 sumi <- sumi + sumj*y1probs[i]
 }
sumi
}
```

Segments ① and ② contain the definition of the function mse. *The input to* mse *consists of the values of x and the support points x_1 and x_2. If the aim is simply to calculate MSE(x), there is no need to input x_1 and x_2, as they can be defined in the global environment. However, when we come to find an IMSE-optimal design, it will be necessary to access the values of x_1 and x_2 from the calling function* imse, *and then it will be necessary to pass x_1 and x_2 to* mse.

Segment ① calculates η_i and then the probability of success, π_i, at each of the two support points. These are used to calculate vectors of the binomial probabilities for each of $0, 1, \ldots, n_i$ successes from n_i trials ($i = 1, 2$). The true probability of success at x, $\pi(x)$, is calculated. In Segment ②, the two loops consider, for each possible value of (y_1, y_2), the calculation of β_0^* and β_1^*, and then $\pi^*(x, y_1, y_2)$ and $[\pi^*(x, y_1, y_2) - \pi(x)]^2$. Finally MSE($x$) is found from (4.25). The vectors a1star *and* a2star *are calculated externally of* mse*; see Segment ③ below.*

③

```
betavec <- c(0,1)
xvalues <- seq(from=-9.21024,to=9.21024,length=101)
x1 <- -1.5434
x2 <- 1.5434

n1 <- 6
n2 <- 6

y1vals <- 0:n1
y2vals <- 0:n2
a1star <- log((n1-y1vals+0.5)/(y1vals+0.5))
a2star <- log((n2-y2vals+0.5)/(y2vals+0.5))
```

Segment ③ defines $\boldsymbol{\beta}$, the support points x_1 and x_2 and the values of x at which

$MSE(x)$ will be evaluated. The values of n_1 and n_2 are declared, and

$$A_1^* = \ln\left[\frac{(n_1 - y_1 + 0.5)}{(y_1 + 0.5)}\right] \quad and \quad A_2^* = \ln\left[\frac{(n_2 - y_2 + 0.5)}{(y_2 + 0.5)}\right] \tag{4.27}$$

$(y_1 = 0, \ldots, n_1;\ y_2 = 0, \ldots, n_2)$ are evaluated for use in calculating β_0^* and β_1^* via (4.22) and (4.23).

④

```
y <- rep(0,101)
i <- 1
for (x in xvalues)
{
 y[i] <- mse(x,x1,x2)
 i <- i+1
}
```

Segment ④ calculates the vector y that contains the values of the MSE that were used to draw the curve for $n_1 = 6$ in Figure 4.9.

Suppose that it is desired to consider only support points x_1 and x_2 satisfying $-5 \leq x_1, x_2 \leq 5$. Looking ahead to the use of optim to find an IMSE-optimal design, this suggests that we might choose an initial solution (z_1, z_2) where z_i lies between 0 and 1, and then use the transformation $x_i = 5\cos(\pi z_i)$ to ensure that the constraints on x_1 and x_2 are met. If Segments ① – ③ of the program beginning on page 138 have already been run in the current R session, the following program will calculate the IMSE:

①

```
imse <- function(xvec)
{#Calculates the IMSE when the support points are at xvec[1] and xvec[
 xvec2 <- 5*cos(pi*xvec)
 x1 <- xvec2[1]
 x2 <- xvec2[2]
```

②

```
 msevalues <- rep(0,(nsteps+1))
 for (k in 1:(nsteps+1))
 {
  msevalues[k] <- mse(xvalues[k],x1,x2)
 }
 summse <- sum(multiplier*msevalues)
 integral <- (h/3)*summse
 integral
}
```

③

```
nsubintervals <- 50
nsteps <- 2*nsubintervals
h <- (9.21024 - (-9.21024))/nsteps
```

```
xvalues <- seq(from=-9.21024,to=9.21024,length=(nsteps+1))
temp <- 2 + 2*(((1:(nsteps-1)) %% 2) == 1)
multiplier <- c(1,temp,1)

initial <- acos(c(-1.5434,1.5434)/5)/pi
out <- imse(initial)
out
```

Segments ① and ② of the program define the function imse. *Segment ③ sets up the use of Simpson's rule from* (2.8) *to calculate the integral. There will be* $n = 50$ *subintervals. The value of* h *is calculated. The vector* multiplier *contains the coefficients (1, 2 or 4) of each MSE(x). Finally, because the first command of* imse *will transform the input vector so as to meet the constraints* $-5 \leq x_1, x_2 \leq 5$*, the vector* initial *contains back-transformed values of the desired support points.*

This program gives the IMSE to be 0.2063553.

With the functions mse *and* imse *created, it is now straightforward to find an IMSE-optimal design. As before, take* $\boldsymbol{\beta} = (0,1)^\top$, $n_1 = n_2 = 6$*, and let the initial guess of the optimal design be* $(-1.5434, 1.5434)$ *plus or minus small deviations. Under the assumption that all the preceding R commands of this section have already been run in the current workspace, the following program will give the best design found by* optim *from 100 initial values:*

```
nsims <- 100
minimse <- 100
initialz <- acos(c(-1.5434,1.5434)/5)/pi
for(i in 1:nsims)
{
 initialx <- initialz + 0.1*(runif(2)-0.5)
 out <- optim(initialx,imse,NULL,method="Nelder-Mead")
 if(out$value < minimse) {minimse <- out$value
    bestdesign <- out$par}
}
answer <-  bestdesign
xvals <- 5*cos(pi*answer)
minimse
xvals
```

Two runs of this program gave very slightly different answers:

```
> minimse
[1] 0.1034249
> xvals
[1] -3.301534  3.301550
```

and

```
> minimse
[1] 0.1034249
> xvals
[1] -3.301336  3.301339
```

The calculated minimum value of the IMSE is the same. If you require so much accuracy that you cannot accept some doubt about whether to use $x_1 = -3.302$ and $x_2 = 3.302$ or $x_1 = -3.301$ and $x_2 = 3.301$, my suggestion would be to increase the number of sub-intervals that you use in approximating the integral in (4.26) by Simpson's rule. This may give you a more accurate estimate of the IMSE, and let you find the IMSE-optimal design more accurately.

The preceding optimisation found an IMSE-optimal design for the canonical variable $z = \beta_0 + \beta_1 x$ (or, equivalently, for $\boldsymbol{\beta} = (0,1)^\top$). Russell et al. (2009a) showed that the locally IMSE-optimal design for other values of $\boldsymbol{\beta}$ could be obtained in terms of the optimal design for z. For $n_1 = n_2 = 6$, the locally IMSE-optimal design has support points

$$\frac{-3.301 - \beta_0}{\beta_1} \quad \text{and} \quad \frac{3.301 - \beta_0}{\beta_1}.$$

Russell et al. (2009a, Table 1) give IMSE-optimal designs for the canonical variable for various values of $n_1 = n_2$ between 4 and 100. These may serve as a guide for IMSE-optimal designs for other values of $n_1 = n_2$, or you may use Program_14 for quick calculations.

There is also a relationship between the minimised values of the IMSE for the IMSE-optimal designs for the canonical variable z and for the variable x. If M is the minimum value of the IMSE for the canonical variable, the minimum value of the IMSE for x will be $M/|\beta_1|$.

This sub-section has considered only the case $n_1 = n_2$. Russell et al. (2009a, page 90) concluded from various investigations that "a safe general rule is to make n_1 and n_2 as equal as possible."

4.8.3 Speeding up the calculations

Determining an IMSE-optimal design is computationally intensive. In the example of Sub-section 4.8.2, each calculation of the IMSE required the function *mse* to be called 101 times. The two loops in Segment ② of the program on page 139 slow down the program considerably. Segment ② may be replaced by the following commands:

```
 mat1 <- outer(a1star*x1,a2star*x2,"-")
 mat2 <- outer((-a1star),a2star,"+")
 pistarxmat <- 1/(1+exp((-mat1 - mat2*x)/distx))
 mse <- t(y1probs)%*%((pistarxmat-pix)^2)%*%y2probs
 mse
}
```

Recall that y_i may take the values $0, \ldots, n_i$, and note the definitions of A_1 and A_2 given in (4.27). In the above R commands, *mat1* and *mat2* are each $(n_1+1)\times(n_2+1)$ matrices. The (i,j) elements of *mat1* and *mat2* are, respectively,

$$x_1 \ln[(j+0.5)/(n_2-j+0.5)] - x_2 \ln[(i+0.5)/(n_1-i+0.5)]$$

and

$$\ln[(i+0.5)/(n_1-i+0.5)] - \ln[(j+0.5)/(n_2-j+0.5)],$$

or $(x_1-x_2)\beta_0^*(i,j)$ and $(x_1 - x_2)\beta_1^*(i,j)$, respectively, for $i = 0, \ldots, n_1$ and $j = 0, \ldots, n_2$.

The (i,j) element of the matrix produced by the command

`pistarxmat <- 1/(1+exp((-mat1 - mat2*x)/distx))`
is $\pi^*(i,j)$ $(i = 0,\ldots,n_1;\ j = 0,\ldots,n_2)$, while the command
`mse <- t(y1probs)%*%((pistarxmat-pix)^2)%*%y2probs`
performs the sum in (4.25).

The function, beginning on page 140, to calculate the IMSE may also be speeded up. Segment ② may be replaced by

```
msevalues <- sapply(xvalues,mse2,x1=x1,x2=x2)
summse <- sum(multiplier*msevalues)
integral <- (h/3)*summse
integral
}
```

The instruction `sapply(xvalues,mse2,x1=x1,x2=x2)` in

```
msevalues <- sapply(xvalues,mse,x1=x1,x2=x2)
```

applies *mse* to each value of x in *xvalues* while keeping *x1* and *x2* at their present values.

Program_14 in the online resources contains the speeded-up version of the programs used in this section to find an IMSE-optimal design. It contains additional features that will automatically calculate the two limits of integration in the integral on the RHS of (4.26). You need to specify the values of $\boldsymbol{\beta}$, n_1 and n_2 and *nsubintervals* (the number of sub-intervals used in Simpson's rule). If you wish to change the design region from $\mathcal{X} = \{-5 \le x \le 5\}$, it will be necessary to change (i) the first line of *imse*, (ii) the calculation of the initial values entered into *optim*, and (iii) the conversion of the final solution from z-values to x-values.

4.8.4 Extending the previous work on IMSE-optimality

We continue to assume the linear predictor $\eta = \beta_0 + \beta_1 x$ and the logit link. Consider the case of $s\,(>2)$ support points $x_1,\ldots,x_s$. At x_i, let n_i independent observations be taken from a Bernoulli distribution with probability $\pi_i = 1/[1+\exp(-\beta_0 - \beta_1 x_i)]$, and let y_i be the number of successes observed.

As $s > p$, it is not possible to obtain explicit expressions for the MLP estimates of β_0 and β_1. However, for a given outcome $(y_1,\ldots,y_s)$ with $y_i \in \{0,\ldots,n_i\}$ $(i = 1,\ldots,s)$, the function *brglm* will give unique estimates $\beta_0^*(y_1,\ldots,y_s)$ and $\beta_1^*(y_1,\ldots,y_s)$. The probability that this particular outcome occurs is

$$P(Y_1 = y_1) \times \ldots \times P(Y_s = y_s) = \prod_{i=1}^{s} \binom{n_i}{y_i} \pi_i^{y_i} (1-\pi_i)^{n_i - y_i}.$$

The probability of success at an arbitrary point x is

$$\pi(x) = 1/[1+\exp(-\beta_0 - \beta_1 x)],$$

while the predicted probability of success at x is

$$\pi^*(x, y_1,\ldots,y_s) = 1/\{1+\exp[-\beta_0^*(y_1,\ldots,y_s) - \beta_1^*(y_1,\ldots,y_s)x]\}.$$

Then

$$\begin{aligned}\text{MSE}(x) =& E\left\{[\pi^*(x,y_1,\ldots,y_s) - \pi(x)]^2\right\} \\ =& \sum_{y_1=0}^{n_1} \cdots \sum_{y_s=0}^{n_s} [\pi^*(x,y_1,\ldots,y_s) - \pi(x)]^2 \prod_{i=1}^{s} \binom{n_i}{y_i} \pi_i^{y_i}(1-\pi_i)^{n_i-y_i}.\end{aligned} \tag{4.28}$$

Repeated calculation of (4.28) is simplified by recognising that the values $y_1, \ldots, y_s$ are not affected by the values of $\boldsymbol{\beta}$, x_1 and x_2, or x. One may set up a matrix whose $(n_1 + 1) \times \ldots \times (n_s + 1)$ rows contain the possible values of $(y_1, \ldots, y_s)$, then calculate and store the corresponding values of $\beta_0^*(y_1, \ldots, y_s)$ and $\beta_1^*(y_1, \ldots, y_s)$ for given $(x_1, \ldots, x_s)$. These are calculated only once for a given set of support points.

The IMSE is calculated exactly as before:

$$\text{IMSE}(x_1, \ldots, x_s) \approx \int_{x_{0.0005}}^{x_{0.9995}} \text{MSE}(x, x_1, \ldots, x_s)\, dx.$$

Then the IMSE-optimal design is found using *optim*.

In R, the only real change from Program_14 is in altering *mse* to deal with an arbitrary number, s, of support points. Program_15 in the online resources contains this more general function *mse*, and is set up to find the IMSE-optimal design for $\boldsymbol{\beta} = (0, 1)^\top$, $s = 4$, $n_1 = \ldots = n_4 = 2$, and *nsubintervals* = 50. To save space, Program_15 is not displayed here. The output from one run of the program was as follows:

```
> minimse
[1] 0.1492024
> solution
[1]  3.178545 -3.178388  3.178408 -3.178674
```

Note that the support points essentially consist of two points, −3.178 and 3.178, with four observations at each, and that the minimum value of the MSE is 0.1492024. This agrees with the IMSE-optimal design for $n_1 = n_2 = 4$ of Russell et al. (2009a, Table 1).

Program_15 was altered to require $s = 3$ support points with $n_1 = 3$, $n_2 = 2$ and $n_3 = 3$, and the result from one particular run was

```
> minimse
[1] 0.1606973
> solution
[1]  2.974777 -3.429114 -3.428742
```

This suggests two support points, −3.429 and 2.975, with $n_1 = 5$ and $n_2 = 3$. However, the minimised value of the IMSE, 0.1606973, is greater than the value of 0.1492024 found above for $n_1 = n_2 = 4$. As the aim is to minimise the IMSE, this supports the remark of Russell et al. (2009a) that it is better to make n_1 and n_2 as equal as possible.

The two examples above suggest that IMSE-optimisation will prefer two support points over more than two. This has not been investigated further. Program_15 could be used to do so. For the same value of N, there will be more $(y_1, \ldots, y_s)$ combinations for $s > 2$ than for $s = 2$. Moreover, using *brglm* to obtain $\boldsymbol{\beta}^*$ is much slower than using exact formulae. So it must be expected that the search for an IMSE-optimal design will take much longer when $s > 2$ than when $s = 2$.

IMSE-optimality can be used for more than one explanatory variable. Except when $s = p$, exact formulae for the elements of $\boldsymbol{\beta}^*$ will not be available, and *brglm* will need to be used. Simpson's rule can be extended fairly simply to more than one variable, so the IMSE can still be calculated, but the time required for computations will increase considerably.

4.9 D_S-optimality

The topic of D_S-optimality was introduced in Sub-section 3.7.5. An application of it to logistic regression is illustrated here.

Suppose that there are $m = 2$ explanatory variables, the logit link is to be used, and we wish to design an experiment to help us choose between the linear predictors $\eta_1 = \beta_0 + \beta_1 x_1 + \beta_2 x_2 + \beta_{12} x_1 x_2$ and $\eta_2 = \beta_0 + \beta_1 x_1 + \beta_2 x_2$. That is, we want to decide whether the crossproduct regressor $x_1 x_2$ is needed in the model, which is equivalent to testing $H_0\colon \beta_{12} = 0$ vs $H_1\colon \beta_{12} \neq 0$. We seek a design that is optimal for testing H_0 vs H_1.

As shown in Sub-section 3.7.5, the full model $\eta_1 = \beta_0 + \beta_1 x_1 + \beta_2 x_2 + \beta_{12} x_1 x_2 = \boldsymbol{f}^\top(\boldsymbol{x})\boldsymbol{\beta}$ may be written as $\eta_1 = f_1^\top(\boldsymbol{x})\boldsymbol{\beta}_1 + f_2^\top(\boldsymbol{x})\boldsymbol{\beta}_2$. Here $f_1^\top(\boldsymbol{x}) = (x_1 x_2)$, $f_2^\top(\boldsymbol{x}) = (1, x_1, x_2)$, $\boldsymbol{\beta}_1 = (\beta_{12})$, $\boldsymbol{\beta}_2 = (\beta_0, \beta_1, \beta_2)^\top$, $p = 4$ and $p_1 = 1$. The information matrix for the full model is

$$\boldsymbol{M}(\xi, \boldsymbol{\beta}) = \sum_{i=1}^{s} \delta_i \, \omega(\boldsymbol{x}_i) \boldsymbol{f}(\boldsymbol{x}_i) \boldsymbol{f}(\boldsymbol{x}_i)^\top,$$

and the other required matrix is

$$\boldsymbol{M}_{22}(\xi, \boldsymbol{\beta}) = \sum_{i=1}^{s} \delta_i \, \omega(\boldsymbol{x}_i) \boldsymbol{f}_2(\boldsymbol{x}_i) \boldsymbol{f}_2(\boldsymbol{x}_i)^\top.$$

Note that the same model weight function, based on the full model, is used for both matrices. Take $\eta_i = \boldsymbol{f}^\top(\boldsymbol{x}_i)\boldsymbol{\beta}$. Then $\omega(\boldsymbol{x}_i) = \exp(\eta_i)/[1 + \exp(\eta_i)]^2$ from (4.3).

The locally D_S-optimal design is that design ξ^* from the design space for which

$$\det\left[\boldsymbol{M}(\xi, \boldsymbol{\beta})\right] / \det\left[\boldsymbol{M}_{22}(\xi, \boldsymbol{\beta})\right]$$

is maximised; see page 73. From page 84,

$$d(\boldsymbol{x}, \xi, \boldsymbol{\beta}) = \omega(\boldsymbol{x}) \left[\boldsymbol{f}^\top(\boldsymbol{x}) \boldsymbol{M}^{-1}(\xi, \boldsymbol{\beta}) \boldsymbol{f}(\boldsymbol{x}) - \boldsymbol{f}_2^\top(\boldsymbol{x}) \boldsymbol{M}_{22}^{-1}(\xi, \boldsymbol{\beta}) \boldsymbol{f}_2(\boldsymbol{x})\right]$$

is the standardised variance of the design ξ. By the general equivalence theorem, if $d(\boldsymbol{x}, \xi^*, \boldsymbol{\beta}) = p_1$ (the number of parameters in $\boldsymbol{\beta}_1$) at each support point of ξ^* (and possibly at other points $\boldsymbol{x} \in \mathcal{X}$) and nowhere on $\mathcal{X}$ is $d(\boldsymbol{x}, \xi^*, \boldsymbol{\beta}) > p_1$, then ξ^* is D_S-optimal.

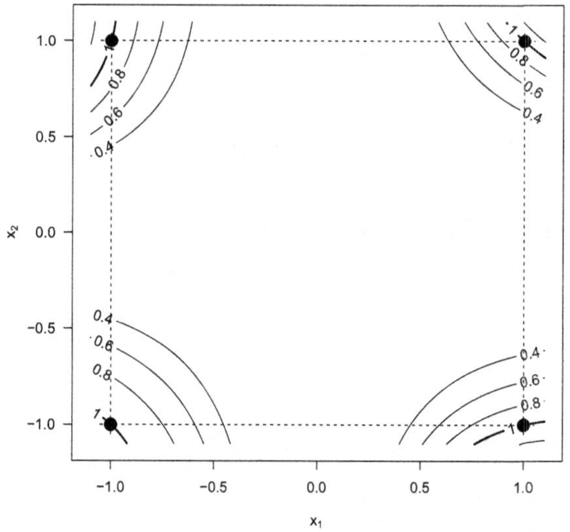

Figure 4.11 *The standardised variance for the locally D_S-optimal design for comparing $\eta = 1 + x_1 + x_2 - x_1x_2$ with $\eta_2 = 1 + x_1 + x_2$, for a logit link.*

The example considered is $\eta = 1+x_1+x_2-x_1x_2$ and $\eta_2 = 1+x_1+x_2$. A logit link was used. The R program used to find the locally D_S-optimal design is *Program_16* in the online resources. It first defines functions to calculate $\boldsymbol{f}(\boldsymbol{x})$ and $\boldsymbol{f}_2(\boldsymbol{x})$, and then defines the function *ratiodets*, which calculates the ratio $-\det\left[\boldsymbol{M}(\xi,\boldsymbol{\beta})\right]/\det\left[\boldsymbol{M}_{22}(\xi,\boldsymbol{\beta})\right]$ that is to be minimised. Then the values of p, s, m and $\boldsymbol{\beta}$ (called *betavec*) are specified. After that, the program simulates starting values of the support points and design weights to enter into *optim*, and the results of the various simulations are compared to find the optimal one. For this example, the support points of the allegedly D_S-optimal design, ξ^*, were found to be $(-1,-1)^\top$, $(-1,1)^\top$, $(1,-1)^\top$ and $(1,1)^\top$, with design weights $\delta_1 = \ldots = \delta_4 = 0.25$.

The remainder of *Program_16* defines a function to calculate $d(\boldsymbol{x},\xi,\boldsymbol{\beta})$ for a value of $\boldsymbol{x}$, then calculates $d(\boldsymbol{x},\xi,\boldsymbol{\beta})$ at each support point of ξ^* and draws a contour plot of the values of $d(\boldsymbol{x},\xi,\boldsymbol{\beta})$ over the design region. Figure 4.11 displays this contour plot. It can be seen that $d(\boldsymbol{x},\xi,\boldsymbol{\beta}) = p_1 = 1$ at each support point of ξ^* and does not exceed one anywhere on the design region, so ξ^* is indeed the locally D_S-optimal design for the problem considered here.

The locally D_S-optimal design considered in Example 3.9.2 did not depend on the values of the parameters in $\boldsymbol{\beta}$, as the model weights $\omega(\boldsymbol{x}_i)$ for a normal distribution do not depend on $\boldsymbol{\beta}$. However, in the example of this sub-section, the weights do depend on β. One might ask by how much small changes in the values of the parameters affect the locally D_S-optimal design. In other words, how *robust* is the design to small changes in values of the parameters? To investigate this, slight changes were made to the value of β_{12} or to β_1, and the locally D_S-optimal designs were sought. It was found that the support points remained unchanged, but that the design weights did vary. Table 4.5 shows the design weights for the locally D_S-optimal designs that examine the term x_1x_2 when $\eta(\boldsymbol{x}) = \boldsymbol{f}^\top(\boldsymbol{x})\boldsymbol{\beta}$ and the values of $\boldsymbol{\beta}$ are as given in the table.

	Support points			
	$(-1,-1)^\top$	$(-1,1)^\top$	$(1,-1)^\top$	$(1,1)^\top$
$\boldsymbol{\beta}^\top$	δ_1	δ_2	δ_3	δ_4
$(1,1,1,-1.2)$	0.259	0.259	0.259	0.223
$(1,1,1,-1.1)$	0.254	0.255	0.255	0.236
$(1,1,1,-1.0)$	0.250	0.250	0.250	0.250
$(1,1,1,-0.9)$	0.246	0.245	0.245	0.264
$(1,1,1,-0.8)$	0.240	0.240	0.241	0.279
$(1,0.9,1,-1)$	0.245	0.265	0.245	0.245
$(1,1.1,1,-1)$	0.255	0.235	0.255	0.255

Table 4.5 *Locally D_S-optimal designs for comparing $\eta_1 = \beta_0 + \beta_1x_1 + \beta_2x_2 + \beta_{12}x_1x_2$ and $\eta_2 = \beta_0 + \beta_1x_1 + \beta_2x_2$ when the full matrix $\boldsymbol{\beta} = (\beta_0, \beta_1, \beta_2, \beta_{12})^\top$ is as given. The various values of $\boldsymbol{\beta}$ represent* $(1,1,1,-1)^\top$ *and minor changes in one element of the vector.*

The changes in the locally D_S-optimal designs are minor. Nonetheless, there are changes. This was a simple example, with only minor changes in the value of $\boldsymbol{\beta}$, so one might not expect large changes in the design. However, should you be considering a locally D_S-optimal design, be alert to the need to examine how the design might vary should the estimate of $\boldsymbol{\beta}$ be in error.

Note that, although $p = s$, the design weights for a design in Table 4.5 are mostly not equal to $1/p$. This contradicts the result in Sub-section 3.7.4 that, when $s = p$, $\delta_1 = \ldots = \delta_s = 1/p$. However, that result applies to D-optimality. There is no reason for the result to apply to D_S-optimality.

4.10 Uncertainty over aspects of the model

In all the preceding work, it has been assumed that the link function $g(\cdot)$, the explanatory variables and the link function to be used are all known. Unfortunately, this often does not reflect the actual situation. Woods et al. (2006) proposed a method "for finding exact designs ... that uses a criterion allowing for uncertainty in the link function, the linear predictor, or the model parameters, together with a design search" (from the Abstract). The details are beyond the scope of this book. Dror & Steinberg (2006) provided an alternative approach to a very similar problem. In this chapter one might wish to choose a design that can cater for uncertainty over which of the logit, probit or complementary log-log link functions is the most appropriate for a particular Bernoulli random variable. Chapter 7 considers the situation where the link function and form of the linear predictor are considered known, but there is uncertainty about the values of the components of the parameter vector $\boldsymbol{\beta}$.

Chapter 5

The Poisson Distribution

5.1 Introduction

In this chapter, experimental designs are considered for the situation where the data are thought to come from Poisson distributions. The chapter starts with a consideration of the Poisson distribution, and then examines means of designing experiments. A theorem is given that tells how a locally D-optimal design can be obtained for many parameter sets. For such parameter sets, designs can be obtained directly without having to perform any numerical optimisation. This is a distinct advantage. When the postulated parameter vector does not meet the requirements of the theorem, numerical optimisation is required, and several examples are given of this. The problem of separation that can arise for small designs can occur for the Poisson distribution as well as the binomial distribution, and the use of MPL estimators to get around this problem is described. Only D-optimality and IMSE-optimality are considered.

5.2 Modelling the Poisson distribution

If one counts the number of events that occur in a specified length, or area, or volume, or interval of time, then the variability in the value of the count is often modelled by a Poisson distribution. Standard examples include the number of welding faults in a fixed length of pipeline, the number of representatives of a species of plant in the fixed area of a randomly tossed quadrat, the number of particles in a fixed volume of a fluid, and the number of accidents that occur in a fixed length of time on a certain stretch of highway. Rather than regularly write about lengths, areas, volumes or intervals of time, I will just consider intervals of time, but this is done solely to simplify the discussion, rather than to suggest that the other measurements are not of relevance.

An assumption underlying the Poisson distribution is that there is a fixed *rate* at which events occur (e.g., 5 per hour). This rate is frequently denoted by λ. Other assumptions are that the probability of an event occurring in a period of time is proportional to the length of the time interval, and that the number of events that occur in any interval is independent of the number of events that occur in any other nonoverlapping interval.

Let Y be the number of events that occur in a time period of fixed length t_0 (> 0). Then λt_0 events are expected in that time and, if Y has a Poisson distribution, then its probability function is

$$P(Y = y) = e^{-\lambda t_0} \frac{(\lambda t_0)^y}{y!}, \quad y = 0, 1, 2, \ldots .$$

A property of the Poisson distribution is that $\mu = \mathrm{E}(Y) = \lambda t_0 = \mathrm{var}(Y)$.

Often the circumstances in which different observations are taken on the process of interest are not identical. That is, the values of some potential explanatory variables may alter, and it is desired to model the value of λ for a particular situation in terms of those explanatory variables. In such circumstances, a GLM, with the natural logarithm as the link function, is generally used to model the mean.

As the mean is λt_0, and $g(\mu) = \ln(\mu) = \ln(\lambda t_0) = \ln(\lambda) + \ln(t_0)$, it is common to set $\ln(t_0)$ as an *offset variable,* which forces the coefficient of $\ln(t_0)$ in the model to be 1, and the rest of the model is used to try to explain the behaviour of $\ln(\lambda)$ in terms of the explanatory variables. If one writes $\ln(\lambda) = \eta = \boldsymbol{f}^\top(\boldsymbol{x})\boldsymbol{\beta}$, then $\ln(\lambda)$ is being modelled by a linear combination of parameters, as in all the GLMs considered so far, and the full model is

$$\ln(\mu) = \ln(t_0) + \eta = \ln(t_0) + \boldsymbol{f}^\top(\boldsymbol{x})\boldsymbol{\beta}. \tag{5.1}$$

Discussion on fitting a GLM with a log link to data is given in Faraway (2006, Chapter 3).

Consider obtaining a locally D-optimal design for a GLM with the Poisson distribution and a log link. In the model in (5.1), the value of t_0 provides no information about the parameter of interest, λ. From the perspective of designing an experiment, it is sensible to regard the values of t_0 for all observations as being equal, and it is most convenient to take $t_0 = 1$. As $\ln(1) = 0$, this gives

$$\ln(\mu) = \eta = \boldsymbol{f}^\top(\boldsymbol{x})\boldsymbol{\beta}. \tag{5.2}$$

The optimal experimental design will have s support points, represented by the $m \times 1$ vectors $\boldsymbol{x}_1, \ldots, \boldsymbol{x}_s$. In this chapter, at the ith support point, a response variable Y_{ij} $(j = 1, \ldots, n_i)$ is thought to have a Poisson distribution with rate λ_i and a known period of observation 1 unit.

As $\mu_i = \mathrm{E}(Y_{ij}) = \lambda_i = \mathrm{var}(Y_{ij})$, then the relationship $\mathrm{var}(Y_{ij}) = \phi V(\mu_i)$ implies that $\phi = 1$ and $V(\mu) = \mu$.

We model $g(\mu_i) = \ln(\mu_i)$ in terms of the explanatory variables in $\boldsymbol{x}_i$ by means of a linear combination of parameters; that is, we have

$$g(\mu_i) = \ln(\mu_i) = \eta_i = \boldsymbol{f}^\top(\boldsymbol{x}_i)\boldsymbol{\beta}.$$

As in other chapters, the aim is to select (i) the support points from a set, $\mathcal{X} \subset \mathbb{R}^m$, of possible points, and (ii) the associated design weights $\delta_1, \ldots, \delta_s$, so that the design

$$\xi = \left\{ \begin{array}{cccc} \boldsymbol{x}_1 & \boldsymbol{x}_2 & \ldots & \boldsymbol{x}_s \\ \delta_1 & \delta_2 & \ldots & \delta_s \end{array} \right\} \tag{5.3}$$

is locally D-optimal.

From (5.2), it follows that

$$\mu_i = \exp(\eta_i)$$

and hence

$$\frac{\partial \mu_i}{\partial \eta_i} = \exp(\eta_i) = \mu_i.$$

Hence the model weights

$$\omega(\boldsymbol{x}_i) = \frac{1}{\phi V(\mu_i)} \left(\frac{\partial \mu_i}{\partial \eta_i} \right)^2, \quad i = 1, \ldots, s,$$

are given by

$$\omega_i = \frac{1}{1 \times \mu_i} (\mu_i)^2 = \mu_i, \quad i = 1, \ldots, s.$$

5.3 Finding D-optimal designs

5.3.1 *The design region*

In Chapter 4, the models $\eta = \beta_0 + \beta_1 x_1$ and $\eta = \beta_0 + \beta_1 x_1 + \cdots + \beta_m x_m$ (for $m > 1$) were considered separately, as it was necessary to impose bounds upon the values of each of the explanatory variables $x_1, \ldots, x_m$ for $m > 1$. In the present situation, where the design weights satisfy $\omega_i = \mu_i = \exp(\eta_i)$, then even when $m = 1$, it is possible for η to increase without bound, meaning that the model weights can approach ∞ much faster that η does. This can cause difficulty with the calculation of an information matrix and its determinant. To prevent this difficulty from occurring, bounds are imposed upon each explanatory variable (even when there is only one of them), and we look for optimal designs on the resulting region.

The good news is that, *in many cases, the locally D-optimal design for a Poisson model can be calculated directly from a simple formula.*

5.3.2 *The model $\eta = \beta_0 + \beta_1 x_1 + \cdots + \beta_m x_m$*

In order to obtain the necessary result, again consider canonical variables (see page 75 for their first mention).

Define $z_1 = \beta_0 + \beta_1 x_1$, $z_2 = \beta_2 x_2$, ... , $z_m = \beta_m x_m$. Then clearly $\eta = \beta_0 + \beta_1 x_1 + \beta_2 x_2 + \cdots + \beta_m x_m = z_1 + \cdots + z_m$. As well, one may write $\boldsymbol{f}(\boldsymbol{z}) = \boldsymbol{B}\boldsymbol{f}(\boldsymbol{x})$, where

$$\boldsymbol{f}(\boldsymbol{z}) = \begin{bmatrix} 1 \\ z_1 \\ z_2 \\ \vdots \\ z_m \end{bmatrix}, \boldsymbol{B} = \begin{bmatrix} 1 & 0 & 0 & \ldots & 0 \\ \beta_0 & \beta_1 & 0 & \ldots & 0 \\ 0 & 0 & \beta_2 & \ldots & 0 \\ \vdots & \vdots & \vdots & \ddots & \vdots \\ 0 & 0 & 0 & \ldots & \beta_m \end{bmatrix} \text{ and } \boldsymbol{f}(\boldsymbol{x}) = \begin{bmatrix} 1 \\ x_1 \\ x_2 \\ \vdots \\ x_m \end{bmatrix}.$$

It can be shown that $\det(\boldsymbol{B}) = \beta_1 \times \beta_2 \times \cdots \times \beta_m$, which is nonzero provided that none of the β_i is equal to 0. This is a reasonable assumption: if you think that β_i equals zero, why would you include the variable x_i in the model? If none of the β_i is equal to 0, then $\det(\boldsymbol{B}) \neq 0$, and therefore the inverse of $\boldsymbol{B}$, $\boldsymbol{B}^{-1}$, exists. Then $\boldsymbol{f}(\boldsymbol{z}) = \boldsymbol{B}\boldsymbol{f}(\boldsymbol{x})$ implies that $\boldsymbol{f}(\boldsymbol{x}) = \boldsymbol{B}^{-1}\boldsymbol{f}(\boldsymbol{z})$.

The information matrix for the design ξ in (5.3) is

$$\begin{aligned}\boldsymbol{M}(\xi, \boldsymbol{\beta}) &= \sum_{i=1}^{s} \delta_i \, \omega(\boldsymbol{x}_i)\boldsymbol{f}(\boldsymbol{x}_i)\boldsymbol{f}^\top(\boldsymbol{x}_i) \\ &= \sum_{i=1}^{s} \delta_i \exp(\beta_0 + \beta_1 x_{i1} + \cdots + \beta_m x_{im})\boldsymbol{f}(\boldsymbol{x}_i)\boldsymbol{f}^\top(\boldsymbol{x}_i). \quad (5.4)\end{aligned}$$

Let the canonical transformation transform the region $\mathcal{X}$ to the region $\mathcal{Z}$, and let

$$\xi_z = \left\{ \begin{array}{cccc} \boldsymbol{z}_1 & \boldsymbol{z}_2 & \ldots & \boldsymbol{z}_s \\ \delta_1 & \delta_2 & \ldots & \delta_s \end{array} \right\} \quad (5.5)$$

be the transformation of ξ in (5.3). As a result of applying the canonical transformation, the information matrix for ξ_z does not depend on $\boldsymbol{\beta}$. Denote this matrix by $\boldsymbol{M}(\xi_z)$. Applying the transformation to (5.4), one obtains

$$\begin{aligned}\boldsymbol{M}(\xi, \boldsymbol{\beta}) &= \sum_{i=1}^{s} \delta_i \exp(z_{i1} + z_{i2} + \cdots + z_{im})\boldsymbol{B}^{-1}\boldsymbol{f}(\boldsymbol{z})[\boldsymbol{B}^{-1}\boldsymbol{f}(\boldsymbol{z})]^\top \\ &= \boldsymbol{B}^{-1}\left[\sum_{i=1}^{s} \delta_i \exp(z_{i1} + z_{i2} + \cdots + z_{im})\boldsymbol{f}(\boldsymbol{z})\boldsymbol{f}^\top(\boldsymbol{z})\right]\left(\boldsymbol{B}^{-1}\right)^\top \\ &= \boldsymbol{B}^{-1}\boldsymbol{M}(\xi_z)\left(\boldsymbol{B}^{-1}\right)^\top. \quad (5.6)\end{aligned}$$

Taking the determinants of both sides of (5.6) gives

$$\begin{aligned}\det[\boldsymbol{M}(\xi, \boldsymbol{\beta})] &= \det\left[\boldsymbol{B}^{-1}\boldsymbol{M}(\xi_z)\left(\boldsymbol{B}^{-1}\right)^\top\right] \\ &= \det\left(\boldsymbol{B}^{-1}\right)\det[\boldsymbol{M}(\xi_z)]\det\left[\left(\boldsymbol{B}^{-1}\right)^\top\right] \\ &= [\det(\boldsymbol{B})]^{-2}\det[\boldsymbol{M}(\xi_z)] \quad (5.7)\end{aligned}$$

As $\det(\boldsymbol{B}) = \beta_1 \times \cdots \times \beta_m$, then $\det(\boldsymbol{B})$ does not vary over $\mathcal{X}$ or $\mathcal{Z}$. Hence, from (5.7), $\det[\boldsymbol{M}(\xi, \boldsymbol{\beta})]$ is a constant multiple of $\det[\boldsymbol{M}(\xi_z)]$, and so the value of ξ for which $\det[\boldsymbol{M}(\xi, \boldsymbol{\beta})]$ is maximised over $\mathcal{X}$ corresponds to the value of ξ_z for which $\det[\boldsymbol{M}(\xi_z)]$ is maximised over $\mathcal{Z}$.

So a globally D-optimal design ξ_z^* over $\mathcal{Z}$ can be found, and the design ξ^* corresponding to ξ_z under the inverse of the canonical transformation will be a locally D-optimal design over $\mathcal{X}$ for the parameter vector $\boldsymbol{\beta}$.

In many circumstances, the following Lemma by Russell et al. (2009b) gives the globally D-optimal design over a defined space $\mathcal{Z}$. Let $\boldsymbol{e}_j$ be the jth column vector of the $m \times m$ identity matrix, $j = 1, \ldots, m$.

Lemma 5.1. *Let* $\eta_i = \boldsymbol{f}(\boldsymbol{x})^\top \boldsymbol{\beta} = [\boldsymbol{B}^{-1}\boldsymbol{f}(\boldsymbol{z})]^\top \boldsymbol{\beta} = z_{i1} + \cdots + z_{im}$, *where* $a_j \leq z_{ij} \leq b_j$ *for* a_j *and* b_j *being constants and* $b_j - a_j \geq 2$ $(j = 1, \ldots, m)$. *A globally D-optimal design for the canonical model* $\eta = z_1 + \cdots + z_m$ *has* $s = p = m + 1$ *support points and is given by*

$$\psi^* = \left\{ \begin{array}{cccc} \boldsymbol{z}_1^* & \boldsymbol{z}_2^* & \cdots & \boldsymbol{z}_{m+1}^* \\ \frac{1}{p} & \frac{1}{p} & \cdots & \frac{1}{p} \end{array} \right\}, \tag{5.8}$$

where $\boldsymbol{z}_j^* = \boldsymbol{b} - 2\boldsymbol{e}_j$ $(j = 1, \ldots, m)$ *and* $\boldsymbol{z}_{m+1}^* = \boldsymbol{b}$ *for* $\boldsymbol{b} = (b_1, \ldots, b_m)^\top$.

The proof of Lemma 5.1 is outlined in Russell et al. (2009b) but is beyond the scope of this book. However, the basic concept is very easy to understand. Formulae for each element of the information matrix of ψ^*, $\boldsymbol{M}(\psi^*)$, were obtained and, from these, formulae were obtained for each element of $\boldsymbol{M}^{-1}(\psi^*)$. This enabled a formula to be obtained for the standardised variance, $d(\boldsymbol{z}, \psi^*)$, at a general point $\boldsymbol{z} \in \mathcal{Z}$. Then it was shown that $d(\boldsymbol{z}, \psi^*) = p = (m + 1)$ at each support point of ψ^*, and that $d(\boldsymbol{z}, \psi^*)$ does not exceed p for any $\boldsymbol{z} \in \mathcal{Z}$. By the Generalised Equivalence Theorem, this proves that ψ^* is D-optimal.

Example 5.3.1. *Find a D-optimal design when* $m = 1$ *and* z_1 *may lie between* $a_1 = -2$ *and* $b_1 = 2$.

As $b_1 - a_1 \geq 2$, *Lemma 5.1 may be used to find the D-optimal design. As* $\boldsymbol{b} = b_1 = 2$ *and* $\boldsymbol{e}_1 = 1$, *then* $\boldsymbol{z}_1^* = \boldsymbol{b} - 2\boldsymbol{e}_1 = 0$ *and* $\boldsymbol{z}_2^* = \boldsymbol{b} = b_1 = 2$. *So the D-optimal design is*

$$\psi^* = \left\{ \begin{array}{cc} 0 & 2 \\ 0.5 & 0.5 \end{array} \right\}.$$

Example 5.3.2. *Find a D-optimal design when* $m = 3$, z_1 *may lie between* $a_1 = -2$ *and* $b_1 = 2$, z_2 *may lie between* $a_2 = 0$ *and* $b_2 = 3$, *and* z_3 *may lie between* $a_3 = -1$ *and* $b_3 = 1$.

As $b_i - a_i \geq 2$ *for each of* $i = 1, 2, 3$, *one may use Lemma 5.1 to find the D-optimal design. Now* $\boldsymbol{b} = (b_1, b_2, b_3)^\top = (2, 3, 1)^\top$, *and* $\boldsymbol{e}_1 = (1, 0, 0)^\top$, $\boldsymbol{e}_2 = (0, 1, 0)^\top$ *and* $\boldsymbol{e}_3 = (0, 0, 1)^\top$. *Then* $\boldsymbol{z}_1^* = \boldsymbol{b} - 2\boldsymbol{e}_1 = (0, 3, 1)^\top$, $\boldsymbol{z}_2^* =$

$\boldsymbol{b} - 2\boldsymbol{e}_2 = (2, 1, 1)^\top$, $\boldsymbol{z}_3^* = \boldsymbol{b} - 2\boldsymbol{e}_3 = (2, 3, -1)^\top$ *and* $\boldsymbol{z}_4^* = \boldsymbol{b} = (2, 3, 1)^\top$. *So the D-optimal design is*

$$\psi^* = \left\{ \begin{array}{cccc} (0,3,1)^\top & (2,1,1)^\top & (2,3,-1)^\top & (2,3,1)^\top \\ 0.25 & 0.25 & 0.25 & 0.25 \end{array} \right\}.$$

Example 5.3.3. *Find a D-optimal design when* $m = 3$, z_1 *may lie between* $a_1 = -2$ *and* $b_1 = 2$, z_2 *may lie between* $a_2 = 0$ *and* $b_2 = 3$, *and* z_3 *may lie between* $a_3 = 0$ *and* $b_3 = 1$.

As $b_3 - a_3 \not\geq 2$, *the conditions of Lemma 5.1 are not satisfied, and so the Lemma cannot be used to find the D-optimal design. Obtaining a D-optimal design when the conditions of the Lemma are not satisfied will be considered in Sub-section 5.3.3.*

The values which each canonical variable z_i can take depend on the values that each explanatory variable may take and on the values of the parameters in $\boldsymbol{\beta}$. Clearly these values must be known (or assumed, in the case of $\boldsymbol{\beta}$) in order to calculate the z_i. Having obtained the D-optimal design ψ^*, one must calculate the values of the $\boldsymbol{x}_i$ that correspond to the $\boldsymbol{z}_i$ in order to obtain the locally D-optimal design ξ^* over $\mathcal{X}$. This design is given by the following theorem, from Russell et al. (2009b).

Theorem 5.1. *Let* $\eta_i = \beta_0 + \beta_1 x_{i1} + \cdots + \beta_m x_{im} = \boldsymbol{\beta}^\top \boldsymbol{f}(\boldsymbol{x}_i)$, $\ell_j \leq x_{ji} \leq u_j$ *and* $|\beta_j(u_j - \ell_j)| \geq 2$ $(j = 1, \ldots, m)$. *Then the locally D-optimal design has* $(m+1)$ *support points with equal design weights* $1/(m+1) = 1/p$, *and the support points are given by*

$$\boldsymbol{x}_j^* = \mathbf{c} - \left(\frac{2}{\beta_j}\right) \mathbf{e}_j \ (j = 1, \ldots, m)$$

$$\boldsymbol{x}_{m+1}^* = \mathbf{c},$$

for $\mathbf{c} = (c_1, \ldots, c_m)^\top$, *where* $c_j = u_j$ *if* $\beta_j > 0$ *and* $c_j = \ell_j$ *if* $\beta_j < 0$.

Example 5.3.4. *Suppose that we wish to find a locally D-optimal design for Poisson observations and a logarithmic link when we have* $m = 2$ *explanatory variables,* x_1 *and* x_2, *and it is required that* x_1 *lies between* $\ell_1 = -1$ *and* $u_1 = 1$ *and* x_2 *lies between* $\ell_2 = -1$ *and* $u_2 = 1$. *Let the best guess of* η *be* $\eta = 1 + 2x_1 - 2x_2$; *i.e.,* $\boldsymbol{\beta} = (1, 2, -2)^\top$. *As* $|\beta_1(u_1 - \ell_1)| = |2[1 - (-1)]| = 4$ *and* $|\beta_2(u_2 - \ell_2)| = |(-2)[1 - (-1)]| = 4$, *the explanatory variables satisfy the requirement that* $|\beta_j(u_j - \ell_j)| \geq 2$ $(j = 1, \ldots, m)$, *and so the result of Theorem 5.1 may be used to give the locally D-optimal design.*

First consider the vector $\mathbf{c}$. *As* $\beta_1 = 2 > 0$, *then* $c_1 = u_1 = 1$; $\beta_2 = -2 < 0$ *implies that* $c_2 = \ell_2 = -1$. *So* $\mathbf{c} = (1, -1)^\top$. *As* $\mathbf{e}_1$ *is the first column* *of the* 2×2 *identity matrix, then* $\mathbf{e}_1 = (1,0)^\top$*; similarly,* $\mathbf{e}_2 = (0,1)^\top$*. Hence the support points of the design are* $\boldsymbol{x}_1^* = \mathbf{c} - (2/\beta_1)\mathbf{e}_1 = \mathbf{c} - \mathbf{e}_1 = (0,-1)^\top$*,* $\boldsymbol{x}_2^* = \mathbf{c} - (2/\beta_2)\mathbf{e}_2 = \mathbf{c} + \mathbf{e}_2 = (1,0)^\top$*, and* $\boldsymbol{x}_3^* = (1,-1)^\top$*. That is, the locally D-optimal design is*

$$\xi^* = \left\{ \begin{array}{ccc} (0,-1)^\top & (1,0)^\top & (1,-1)^\top \\ \frac{1}{3} & \frac{1}{3} & \frac{1}{3} \end{array} \right\}.$$

A contour plot of the standardised variance for this design appears in Figure 5.1. It is clear that the standardised variance equals 3 at the support points, and does not exceed 3 anywhere on the set $\mathcal{X} = \{(x_1, x_2) : -1 \le x_1 \le 1, -1 \le x_2 \le 1\}$*, which supports the Theorem's statement that this is the locally D-optimal design.*

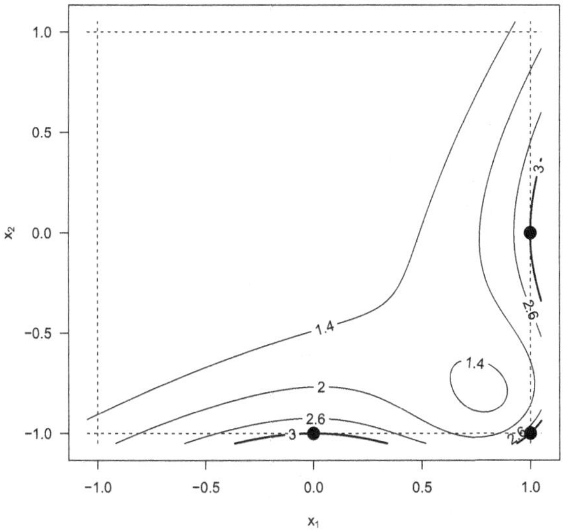

Figure 5.1 *A contour plot for the standardised variance* $d(\boldsymbol{x}, \xi^*, \boldsymbol{\beta})$ *for* ξ^* *given in* (5.3.4) *and* $\boldsymbol{\beta} = (1,2,-2)^\top$.

When $m > 2$, it becomes difficult or impossible to show easily a graph of the values of the standardised variance. So the method (see Program_13 on page 113) of examining the values of the standardised variance in the neighbourhood of each alleged support point is used instead.

The R program *Program_17* constructs the design given by Theorem 5.1, and will print a warning if the set of values permitted for an explanatory variable is not wide enough for the theorem to hold. You need to specify the parameter vector $\boldsymbol{\beta}$ (called *betavec*), the vector of lower bounds *lvec* $= (\ell_1, \ldots, \ell_m)$ and the vector of upper bounds *uvec* $= (u_1, \ldots, u_m)$.

Example 5.3.5. *The following input and output show the result of the construction for* $m = 4$ *when each of* $x_1, \ldots, x_4$ *is restricted to lie between -1 and 1, and the parameter vector is* $\boldsymbol{\beta} = (1, 2, 1, -1, -2)^\top$.

```
> m <- 4
> betavec <- c(1,2,1,-1,-2)
> lvec <- c(-1,-1,-1,-1)
> uvec <- c(1,1,1,1)
> construct_Poisson_Dopt()
                weights
1 0  1 -1 -1      0.2
2 1 -1 -1 -1      0.2
3 1  1  1 -1      0.2
4 1  1 -1  0      0.2
5 1  1 -1 -1      0.2
```

That is, the locally D-optimal design is

$$\left\{ \begin{matrix} (0,1,-1,-1)^\top & (1,-1,-1,-1)^\top & (1,1,1,-1)^\top & (1,1,-1,0)^\top & (1,1,-1,-1)^\top \\ 0.2 & 0.2 & 0.2 & 0.2 & 0.2 \end{matrix} \right\}.$$

As this design has been found using the theorem, it is not necessary to check that the standardised variance achieves a maximum of $p = 5$ *at each support point, as the proof of the theorem has established this.*

5.3.3 When Theorem 5.1 cannot be used

Suppose that you want to construct a locally D-optimal design for m explanatory variables that are all required to lie between -1 and 1 and the conditions of the theorem are not satisfied. An example is when the parameter vector is $\boldsymbol{\beta} = (1, 2, 0.5)^\top$, as $|\beta_2(u_2 - \ell_2)| = |0.5[1 - (-1)]| = 1$ is less than 2. Russell et al. (2009b, Remark 3) consider the canonical model. They state that, *if* the D-optimal design has p support points (which will have equal design weights; see Sub-section 3.7.1), then they are given by the following formula:

$$\boldsymbol{z}_i = \boldsymbol{b} + \max(-2, a_i - b_i)\boldsymbol{e}_i, \; i = 1, \ldots, p-1 \tag{5.9}$$
$$\boldsymbol{z}_p = \boldsymbol{b}, \tag{5.10}$$

where the notation is the same as in Lemma 5.1.

Example 5.3.6. *Consider again the case of* $\boldsymbol{\beta} = (1, 2, 0.5)^\top$, *where*

$-1 \le x_i \le 1$ for $i = 1, 2$. One must find the proposed support points of the locally D-optimal design for $\boldsymbol{\beta}$. Now

$$z_1 = \beta_0 + \beta_1 x_1 = 1 + 2x_1 \Rightarrow -1 \le z_i \le 3 \tag{5.11}$$

$$z_2 = \beta_2 x_2 = 0.5x_2 \Rightarrow -0.5 \le z_2 \le 0.5. \tag{5.12}$$

So $a_1 = -1$, $b_1 = 3$, $a_2 = -0.5$, $b_2 = 0.5$ and $\boldsymbol{z}_3 = \boldsymbol{b} = (b_1, b_2)^\top = (3, 0.5)^\top$. Then $\boldsymbol{z}_1 = \boldsymbol{b} + \max(-2, a_1 - b_1)\boldsymbol{e}_1 = \boldsymbol{b} + \max(-2, -1-3)\boldsymbol{e}_1 = \boldsymbol{b} - 2\boldsymbol{e}_1 = (1, 0.5)^\top$ and $\boldsymbol{z}_2 = \boldsymbol{b} + \max(-2, a_2 - b_2)\boldsymbol{e}_2 = \boldsymbol{b} + \max(-2, -0.5-0.5)\boldsymbol{e}_1 = \boldsymbol{b} - 1\boldsymbol{e}_1 = (3, -0.5)^\top$. Now transforming back from the canonical variables to the original x_1 and x_2, the proposed locally D-optimal design is

$$\xi = \left\{ \begin{array}{ccc} (0,1)^\top & (1,-1)^\top & (1,1)^\top \\ 1/3 & 1/3 & 1/3 \end{array} \right\}.$$

The procedure just followed gives a locally D-optimal design only if there are just $s = 3$ support points in the design. To check that the design is indeed locally D-optimal, the standardised variance $d(\boldsymbol{x}, \xi, \boldsymbol{\beta})$ must be considered for all $\boldsymbol{x} \in \mathcal{X}$. A contour plot of the standardised variance can be obtained in exactly the same way as has been done before, and it shows that the maximum value of $d(\boldsymbol{x}, \xi, \boldsymbol{\beta})$ over $\boldsymbol{x} \in \mathcal{X}$ is 3, and that it occurs at each of the support points. So the design ξ is indeed locally D-optimal over $\mathcal{X}$ for $\boldsymbol{\beta} = (1, 2, 0.5)^\top$.

Example 5.3.7. *Now suppose that there are $m = 3$ explanatory variables, and the assumed value of the parameter vector is $\boldsymbol{\beta} = (1, 0.6, 0.5, 0.4)^\top$. Following the procedure outlined above, the conjectured locally D-optimal design is*

$$\xi = \left\{ \begin{array}{cccc} (-1,1,1)^\top & (1,-1,1)^\top & (1,1,-1)^\top & (1,1,1) \\ 0.25 & 0.25 & 0.25 & 0.25 \end{array} \right\}.$$

It is easy to verify that the standardised variance takes the value 4 at each of the support points, but one must check that the maximum value of $d(\boldsymbol{x}, \xi, \boldsymbol{\beta})$ for any $\boldsymbol{x} \in \mathcal{X}$ is 4.

Applying Program_13 from the Web site doeforglm.com results in the following output:

```
For support point  1
 Maximum std var is  4  at
 -1 1 1
For support point  2
 Maximum std var is  4  at
 1 -1 1
For support point  3
```

```
 Maximum std var is  4  at
 1 1 -1
For support point  4
 Maximum std var is  4  at
 1 1 1
```

This tells us that the maximum value of the standardised variance in the neighbourhood of any support point is $p = 4$*, which is the value taken at each support point. It is reasonable to conclude that we do indeed have the locally D-optimal design.*

If the conjectured design is not locally D-optimal, then it is necessary to find the locally D-optimal design using optimisation. One proceeds exactly as in Chapter 4, except that the model weights are now given by $\omega(\boldsymbol{x}_i) = \exp(\eta_i) = \exp(\boldsymbol{x}_i^\top \boldsymbol{\beta})$.

Russell et al. (2009b, Remark 3) provided the example of $m = 2$ and $\boldsymbol{\beta} = (-0.91, 0.04, -0.69)^\top$. For $\boldsymbol{x} \in \mathcal{X}$, the conjectured locally D-optimal design has the support points $(1,1)^\top$, $(-1,1)^\top$ and $(1,-1)^\top$, each with a design weight of $1/3$. However, it is clear from the contour plot of values of $d(\boldsymbol{x}, \xi, \boldsymbol{\beta})$ in Figure 5.2 that the maximum value of $d(\boldsymbol{x}, \xi, \boldsymbol{\beta})$ over $\mathcal{X}$ is greater than $p = 3$. By the general equivalence theorem, the present design is not locally D-optimal.

Figure 5.2 suggests that another support point is needed in the vicinity of $(-1,-1)^\top$. As there are $p = 3$ parameters in $\boldsymbol{\beta}$, the maximum number of support points is $p(p+1)/2 = 6$. So one could begin a search for an optimal design with $s = 6$ support points and decrease the value of s if this seems appropriate. It is easily verified that, as claimed by Russell et al. (2009b), the locally D-optimal design is

$$\xi = \left\{ \begin{array}{cccc} (1,1)^\top & (1,-1)^\top & (-1,1)^\top & (-1,-1)^\top \\ 0.213 & 0.313 & 0.163 & 0.311 \end{array} \right\}.$$

A contour plot of $d(\boldsymbol{x}, \xi, \boldsymbol{\beta})$ is shown in Figure 5.3. The maximum value of $d(\boldsymbol{x}, \boldsymbol{\beta})$ for $\boldsymbol{x} \in \mathcal{X}$ is $p = 3$, and is achieved at each of the four support points of ξ.

Theorem 5.1 can be applied only when η is of the form $\beta_0 + \beta_1 x_1 + \cdots + \beta_{p-1} x_{p-1}$. Should a more complicated linear predictor (e.g., where some of the regressors are squares, or cross-products, of explanatory variables) be conjectured, one must use constrained optimisation to find a design. The only difference between the R commands required here and those used in Chapter 4 is that the model weight $w(\boldsymbol{x}_i) = \exp(\eta_i)$ must be used.

Example 5.3.8. *For* $m = 2$ *explanatory variables* x_1 *and* x_2*, consider the linear predictor* $\eta = \boldsymbol{f}(\boldsymbol{x})^\top \boldsymbol{\beta} = (1, x_1, x_2, x_1 x_2)(1, -1, 2, -0.5)^\top =$

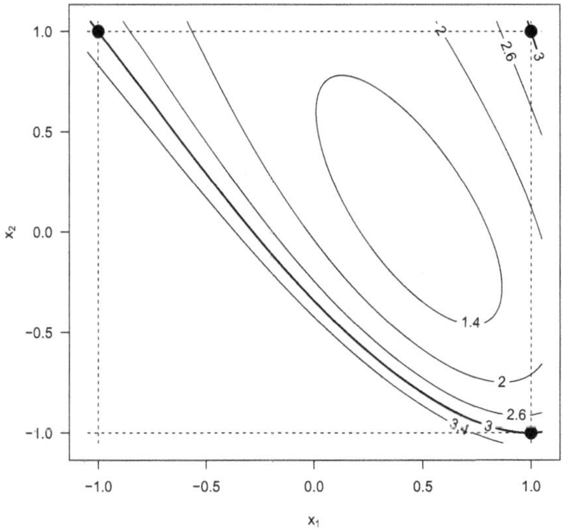

Figure 5.2 *Contour plot of the standardised variance of the incorrectly conjectured locally D-optimal design for* $\boldsymbol{\beta} = (-0.91, 0.04, -0.69)^\top$*. The values of* $d(\boldsymbol{x}, \xi, \boldsymbol{\beta})$ *exceed* $p = 3$ *in the lower left-hand region.*

$1 - x_1 + 2x_2 - 0.5x_1x_2$. *There are* $p = 4$ *parameters, so the maximum required number of support points is* $p(p+1)/2 = 10$. *It required only a couple of runs of the program to reduce* s *to four. Then, "tweaking" the output from one run to give starting values for each simulation of another run quickly led to the design*

$$\xi^* = \left\{ \begin{array}{cccc} (-1.000, 0.200)^\top & (-1.000, 1.000)^\top & (0.334, 1.000)^\top & (1.000, -0.334)^\top \\ 0.25 & 0.25 & 0.25 & 0.25 \end{array} \right\}.$$

The standardised variance achieves its maximum value over the design space of $p = 4$ *at the support points of* ξ^**, as shown in Figure 5.4. It follows from the general equivalence theorem that* ξ^* *is locally D-optimal for a Poisson distribution with log link and* $\eta = 1 - x_1 + 2x_2 - 0.5x_1x_2$.

5.4 Small values of the total sample size, N

Sections 3.6 and 4.8 considered the situation of small sample sizes for data from a Bernoulli distribution. Here, suppose that, at the ith support point $\boldsymbol{x}_i$ $(i = 1, \ldots, s)$, n_i observations are taken from a Poisson

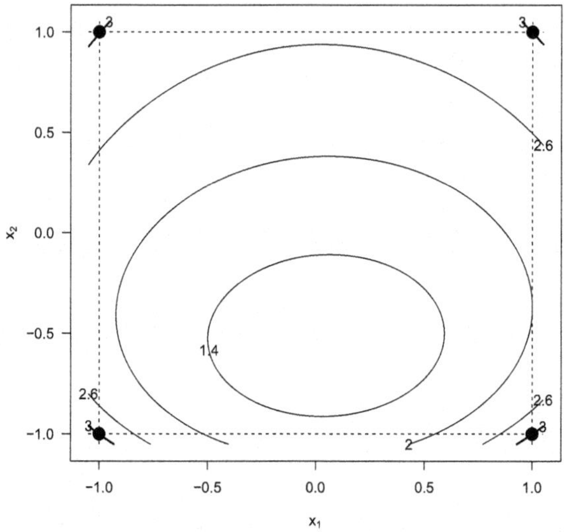

Figure 5.3 *Contour plot of the standardised variance of the locally D-optimal design for* $\boldsymbol{\beta} = (0.91, 0.04, 0.69)^\top$. *The standardised variance achieves its maximum value of* $p = 3$ *at each of the four indicated support points.*

distribution with mean $\lambda_i = \exp(\eta_i) = \exp(\boldsymbol{x}_i^\top \boldsymbol{\beta})$. If y_{ij} denotes the jth observation at $\boldsymbol{x}_i$, then the likelihood of the overall sample is

$$L(\boldsymbol{\beta}; y_{11}, \ldots, y_{sn_s}) = \mathrm{e}^{-\lambda_1} \frac{\lambda_1^{y_{11}}}{y_{11}!} \times \ldots \times \mathrm{e}^{-\lambda_1} \frac{\lambda_1^{y_{1n_1}}}{y_{1n_1}!} \times \ldots \times \mathrm{e}^{-\lambda_s} \frac{\lambda_s^{y_{sn_s}}}{y_{sn_s}!}$$

$$= \exp\left(-\sum_{i=1}^{s} n_i \lambda_i\right) \prod_{i=1}^{s} \lambda_i^{y_{i\cdot}} \Bigg/ \left(\prod_{i=1}^{s} \prod_{j=1}^{n_i} y_{ij}!\right),$$

which implies that the log likelihood, $\ell(\boldsymbol{\beta}; y_{11}, \ldots, y_{sn_s})$, is given by

$$\ell(\boldsymbol{\beta}; y_{11}, \ldots, y_{sn_s}) = -\sum_{i=1}^{s} n_i \lambda_i + \sum_{i=1}^{s} y_{i\cdot} \ln(\lambda_i) - \sum_{i=1}^{s} \sum_{j=1}^{n_i} \ln(y_{ij}!). \quad (5.13)$$

In addition, by (1.21), the (j, k) element of the matrix $\boldsymbol{\mathcal{I}}$ equals

$$\mathcal{I}_{jk} = \sum_{i=1}^{s} n_i \frac{f_{ij} f_{ik}}{\mathrm{var}(Y_i)} \left(\frac{\partial \mu_i}{\partial \eta_i}\right)^2 = \sum_{i=1}^{s} n_i \frac{f_{ij} f_{ik}}{\lambda_i} \lambda_i^2 \quad j, k \in \{0, \ldots, p-1\}. \quad (5.14)$$

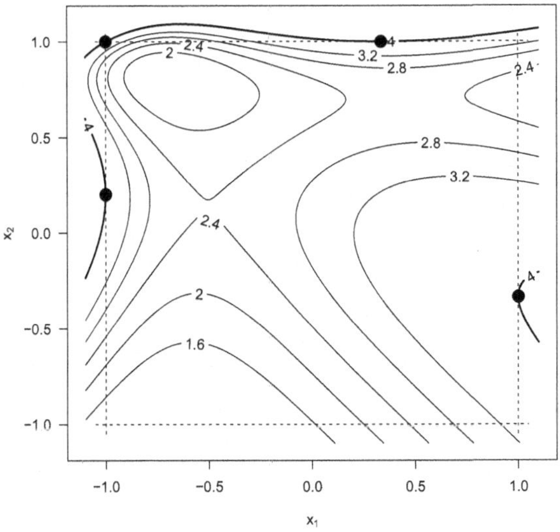

Figure 5.4 *A contour plot of the standardised variance for the design ξ^* of Example 5.3.8. The standardised variance achieves its maximum value of $p = 4$ at each of the four indicated support points.*

Consider the simple two-parameter model $\eta_i = \beta_0 + \beta_1 x_i$, and let Y_{ij} denote the jth observation at $\boldsymbol{x}_i$ $(i = 1, \ldots, s)$. Then

$$\mathcal{I} = \left[\begin{array}{cc} \sum_{i=1}^{s} n_i \lambda_i & \sum_{i=1}^{s} n_i \lambda_i x_i \\ \sum_{i=1}^{s} n_i \lambda_i x_i & \sum_{i=1}^{s} n_i \lambda_i x_i^2 \end{array} \right]. \tag{5.15}$$

The ML estimates of β_0 and β_1 are the values, $\hat{\beta}_0$ and $\hat{\beta}_1$, that maximise the log-likelihood, and which satisfy the equations

$$U_0|_{\boldsymbol{\beta}=\hat{\boldsymbol{\beta}}} = 0 \qquad U_1|_{\boldsymbol{\beta}=\hat{\boldsymbol{\beta}}} = 0,$$

where U_0 and U_1 are the score statistics given in (1.20). Note that $\text{var}(Y_{ij}) = \lambda_i$ and that $(\partial \mu_i)/(\partial \eta_i) = (\partial \lambda_i)/(\partial \eta_i) = \lambda_i$ also.

Let y_1 and y_2 represent the totals of the observations taken at the support points x_1 and x_2, respectively. It follows that $\hat{\beta}_0$ and $\hat{\beta}_1$ satisfy

$$\begin{aligned} [y_1 - n_1 \exp(\hat{\beta}_0 + \hat{\beta}_1 x_1)] + [y_2 - n_2 \exp(\hat{\beta}_0 + \hat{\beta}_1 x_2)] &= 0, \\ x_1[y_1 - n_1 \exp(\hat{\beta}_0 + \hat{\beta}_1 x_1)] + x_2[y_2 - n_2 \exp(\hat{\beta}_0 + \hat{\beta}_1 x_2)] &= 0, \end{aligned}$$

and are therefore equal to

$$\hat{\beta}_0 = [x_1 \ln(y_2/n_2) - x_2 \ln(y_1/n_1)]/(x_1 - x_2) \tag{5.16}$$

$$\hat{\beta}_1 = [\ln(y_1/n_1) - \ln(y_2/n_2)]/(x_1 - x_2). \tag{5.17}$$

These may also be written as

$$\hat{\beta}_0 = [x_1 \ln(\bar{y}_2) - x_2 \ln(\bar{y}_1)]/(x_1 - x_2),$$

$$\hat{\beta}_1 = [\ln(\bar{y}_1) - \ln(\bar{y}_2)]/(x_1 - x_2).$$

These estimates will be undefined if either or both of y_1 and y_2 are zero. As with undefined estimates in a logistic regression, R does not warn of a problem when performing an analysis, but the fact that a problem has occurred is shown by the occurrence of very large estimated standard errors.

Example 5.4.1. *Consider the case where $n_1 = n_2 = 4$, and observations are taken at $x_1 = -1$ and $x_2 = 1$. The results that are obtained give $y_1 = 0$ and $y_2 = 4$. An example of an analysis and its output appears below.*

```
> yvec <- c(0,0,0,0,0,1,1,2)
> xvec <- c(-1,-1,-1,-1,1,1,1,1)
> out <- glm(yvec ~ xvec,family="poisson")
> summary.glm(out)$coeff
             Estimate Std. Error      z value  Pr(>|z|)
(Intercept) -10.15129   3885.466 -0.002612632 0.9979154
xvec         10.15129   3885.466  0.002612632 0.9979154
```

The values of the standard error are both very large, suggesting that there is a problem with the estimation.

The probability that $y_1 = 0$ or $y_2 = 0$ is

$$1 - [1 - \exp(-n_1\lambda_1)][1 - \exp(-n_2\lambda_2)], \tag{5.18}$$

where $\lambda_1 = \exp(\beta_0 + \beta_1 x_1)$ and $\lambda_2 = \exp(\beta_0 + \beta_1 x_2)$. This probability tends to 0 as n_1 and n_2 get larger.

If you are planning to use small samples and only two support points, you should calculate the probability of obtaining undefined estimates early in the design of the experiment.

Example 5.4.2. *Suppose that the estimates of the parameters are $\beta_0 = 1$ and $\beta_1 = 2$, and the design space is $\mathcal{X} = \{x : -1 \leq x \leq 1\}$. The design space for the canonical variable $z = \beta_0 + \beta_1 x = 1 + 2x$ is $\mathcal{Z} = \{z : -1 \leq$*

n	Probability
1	6.599×10^{-2}
2	4.354×10^{-3}
3	2.873×10^{-4}
4	1.896×10^{-5}
5	1.251×10^{-6}

Table 5.1 *Probability of undefined parameter estimates when each support point has n observations made at it for the locally D-optimal design ξ_x^* in* (5.19).

$z \le 3\}$, and so Theorem 5.1 implies that the globally D-optimal design and locally D-optimal designs are, respectively,

$$\xi_z^* = \left\{ \begin{array}{cc} 1 & 3 \\ 0.5 & 0.5 \end{array} \right\} \quad \textit{and} \quad \xi_x^* = \left\{ \begin{array}{cc} 0 & 1 \\ 0.5 & 0.5 \end{array} \right\}. \tag{5.19}$$

Then $\lambda_1 = \exp(1 + 2 \times 0) = e^1$ and $\lambda_2 = \exp(1 + 2 \times 1) = e^3$. As the locally D-optimal design has $\delta_1 = \delta_2$, one would choose $n_1 = n_2$ where possible. Then Table 5.1 shows the value of the probability of undefined estimates, as given in (5.18), *for small values of $n_1 = n_2$.*

In addition,

$$\mathcal{I} = \left[\begin{array}{cc} n_1\lambda_1 + n_2\lambda_2 & n_1\lambda_1 x_1 + n_2\lambda_2 x_2 \\ n_1\lambda_1 x_1 + n_2\lambda_2 x_2 & n_1\lambda_1 x_1^2 + n_2\lambda_2 x_2^2 \end{array} \right],$$

so, from the Result on page 26 for the determinant of a 2×2 matrix,

$$\begin{aligned} \det(\mathcal{I}) &= (n_1\lambda_1 + n_2\lambda_2)(n_1\lambda_1 x_1^2 + n_2\lambda_2 x_2^2) - (n_1\lambda_1 x_1 + n_2\lambda_2 x_2)^2 \\ &= n_1 n_2 \lambda_1 \lambda_2 (x_2 - x_1)^2. \end{aligned} \tag{5.20}$$

As with the binomial distribution (see Section 4.8), we investigate the MPL estimates of the parameters. Recall from (4.21) that the penalised log-likelihood is equal to

$$\ell^*(\boldsymbol{\beta}; y_{11}, \ldots, y_{sn_s}) = \ell(\boldsymbol{\beta}; y_{11}, \ldots, y_{sn_s}) + 0.5 \times \ln[\det(\mathcal{I})].$$

As

$$\begin{aligned} \ln[\det(\mathcal{I})] &= \{\ln[n_1 n_2 (x_1 - x_2)^2] + \ln(\lambda_1) + \ln(\lambda_2)\} \\ &= \ln[n_1 n_2 (x_1 - x_2)^2] + (\beta_0 + \beta_1 x_1) + (\beta_0 + \beta_1 x_2), \end{aligned}$$

then the MPL estimators, β_0^* and β_1^*, satisfy

$$\left.\frac{\partial \ell^*}{\partial \beta_j}\right|_{\boldsymbol{\beta}=\boldsymbol{\beta}^*} = 0, \quad (j = 0, 1),$$

where

$$\left.\frac{\partial \ell^*}{\partial \beta_j}\right|_{\boldsymbol{\beta}=\boldsymbol{\beta}^*} = U_j^*\big|_{\boldsymbol{\beta}=\boldsymbol{\beta}^*} + 0.5 \left.\frac{\partial}{\partial \beta_j}\{K + (\beta_0 + \beta_1 x_1) + (\beta_0 + \beta_1 x_2)\}\right|_{\boldsymbol{\beta}=\boldsymbol{\beta}^*},$$

and $K = \ln[n_1 n_2 (x_1 - x_2)^2]$.

That is,

$$\begin{aligned} [y_1 - n_1\pi^*(x_1)] + [y_2 - n_2\pi^*(x_2)] + 0.5(1+1) &= 0, \\ x_1[y_1 - n_1\pi^*(x_1)] + x_2[y_2 - n_2\pi^*(x_2)] + 0.5(x_1 + x_2) &= 0, \end{aligned}$$

where $\pi^*(x_1) = \exp(\beta_0^* + \beta_1^* x_1)$ and $\pi^*(x_2) = \exp(\beta_0^* + \beta_1^* x_2)$.

Hence the MPL estimators are equal to

$$\beta_0^*(y_1, y_2) = \{x_1 \ln[(y_2 + 0.5)/n_2] - x_2 \ln[(y_1 + 0.5)/n_1]\}/(x_1 - x_2) \tag{5.21}$$

$$\beta_1^*(y_1, y_2) = \{\ln[(y_1 + 0.5)/n_1] - \ln[(y_2 + 0.5)/n_2]\}/(x_1 - x_2), \tag{5.22}$$

where here the dependence of β_0^* and β_1^* on y_1 and y_2 is made explicit.

The MPL estimates differ from the ML estimates in (5.16) and (5.17) only by the addition of 0.5 to each of y_1 and y_2. As in logistic regression, it is impossible for the MPL estimates to be undefined.

Example 5.4.3. *In Example 5.4.1, where $n_1 = n_2 = 4$, $x_1 = -1$ and $x_2 = 1$, and $y_1 = 0$ and $y_2 = 4$, the ML estimates are undefined. From* (5.21) *and* (5.22)*, the MPL estimates are $\beta_0^* = -0.980829$ and $\beta_1^* = 1.098612$.*

If one has only small sample sizes, and the possibility of undefined ML estimates needs to be considered, one should again consider the use of bias reduced estimates of the parameters. While the package `brglm` cannot be used, as it is written only for binomial distributions, at least one other package exists, called `brglm2`. It is based on reducing the bias of the ML estimators, as outlined above, which uses the work of Firth (1993) and Kosmidis & Firth (2009).

The package `brglm2` can be installed in R as simply as `brglm` was. Once it has been installed, you need only type `library(brglm2)` in an R session to load `brglm2` to your workspace.

Example 5.4.4. *Consider again the data from Example 5.4.1, where it was found that separation occurred. Application of* brglm2 *to these data occurs using the following commands, with the output shown also.*

```
> yvec <- c(0,0,0,0,0,1,1,2)
> xvec <- c(-1,-1,-1,-1,1,1,1,1)
> # The maximum likelihood fit with log link
> outML <- glm(yvec ~ xvec,family=poisson(link="log"))
> summary(outML)$coeff
               Estimate Std. Error      z value  Pr(>|z|)
(Intercept) -10.15129   3885.466 -0.002612632 0.9979154
xvec         10.15129   3885.466  0.002612632 0.9979154
> # The bias-reduced fit
> outBR <- update(outML, method = "brglmFit")
> summary(outBR)$coeff
               Estimate Std. Error  z value  Pr(>|z|)
(Intercept) -0.9808293   0.745356 -1.315921 0.1882007
xvec         1.0986123   0.745356  1.473943 0.1404969
```

It can be seen that the “bias reduced” estimates match those obtained from (5.21) and (5.22) in Example 5.4.3.

The estimation of β_0 and β_1 is frequently done in order to estimate the rate $\lambda(x) = \exp[\eta(x)]$ at a given value of x. A measure of how effectively this is done is the MSE of $\lambda^*(x, y_1, y_2) = \exp[\beta_0^*(y_1, y_2) + \beta_1^*(y_1, y_2)x]$. See Section 4.8 for a consideration of this for the probability of success in a Bernoulli distribution. The development below follows that in Section 4.8. We have

$$\begin{aligned} \text{MSE}[\lambda^*(x)] &= E\left\{[\lambda^*(x) - \lambda(x)]^2\right\} \\ &= \sum_{y_1=0}^{\infty}\sum_{y_2=0}^{\infty} [\lambda^*(x, y_1, y_2) - \lambda(x)]^2 \; P(Y_1 = y_1)P(Y_2 = y_2) \\ &= \sum_{y_1=0}^{\infty}\sum_{y_2=0}^{\infty} [\lambda^*(x, y_1, y_2) - \lambda(x)]^2 \; \mathrm{e}^{-\lambda_1}\frac{\lambda_1^{y_1}}{y_1!}\,\mathrm{e}^{-\lambda_2}\frac{\lambda_2^{y_2}}{y_2!}. \end{aligned} \tag{5.23}$$

A complication in evaluating this expression is that there are infinite numbers of values over which to sum for each of y_1 and y_2. A pragmatic approach to this problem will be suggested shortly.

For an appropriate choice of bounds x_{lo} and x_{hi} for the integral that follows, the IMSE is defined as

$$\text{IMSE} = \int_{-\infty}^{\infty} \text{MSE}(x)\,dx \approx \int_{x_{\text{lo}}}^{x_{\text{hi}}} \text{MSE}(x)\,dx. \tag{5.24}$$

For the situation here, the IMSE-optimal design is defined as the values x_1 and x_2 of the support points which, for fixed numbers of observations n_1 and n_2, minimises the value of the IMSE over all potential designs.

Important note: When calculating an IMSE-optimal design with $s = 2$, one can use Equations (5.21) and (5.22) to calculate β_0^* and β_1^*. However, it is not necessary to consider the individual observations $Y_{i1}, \ldots, Y_{in_i}$ at the ith support point x_i ($i = 1, 2$). Instead, one may use a result from most introductory textbooks on mathematical statistics that, if $Y_{i1}, \ldots, Y_{in_i}$ are n_i independent observations from a Poisson distribution with mean λ_i, then the sum $Y_i = Y_{i1} + \cdots + Y_{in_i}$ has a Poisson distribution with mean $n_i \lambda_i$. That is, one may simply consider the various values of the sum Y_i, rather than all the possible sets of values that give this sum. That simplifies the calculations considerably, and is used in Program_18 (described below).

However, this cannot be done (even for $s = 2$) if you use *brglm2* to calculate β_0^* and β_1^*, as there is no way to tell R upon how many observations each sum is based except by giving it the individual observations and letting R count them.

My approach to dealing with sums of the form

$$\sum_{y_i=0}^{\infty} \ldots P(Y_i = y_i) \tag{5.25}$$

is to sum only over those values of y_i for which $P(Y_i = y_i) \geq \ell$, where ℓ is 0.001 or 0.0001. (You may, of course, choose some other lower limit.) I first use the R function `qpois(k, lambda)` which gives the smallest integer y such that $P(Y \leq y) \geq k$ when Y has a Poisson distribution with mean *lambda*. I use this to choose the values of y that lie between `qpois(0.0005, lambda)` and `qpois(0.9995, lambda)` or else between `qpois(0.0001, lambda)` and `qpois(0.9999, lambda)`, and then I select the subset of these values for which $P(Y = y) \geq \ell$.

Example 5.4.5. *The following program finds those values of y for which $P(Y = y) \geq 0.001$ when Y has a Poisson distribution with mean $\lambda = 10$. First it is found that $y_{lo} = 2$ and $y_{hi} = 22$ are the smallest integers satisfying $P(Y \leq y_{lo}) \geq 0.0005$ and $P(Y \leq y_{hi}) \geq 0.9995$. The program then scans the values of $P(Y = y)$ for each $y \in \{2, \ldots, 22\}$ to see which probabilities are at least 0.001, and records those values of y and their corresponding probabilities. The values of y are $2, \ldots, 20$. Consequently, if this situation applied to the sum in (5.25), I would sum from $y = 2$ to $y = 20$ rather than from $y = 0$ to $y = \infty$.*

```
> lambda <- 10
> ylo <- qpois(0.0005,lambda)
> yhi <- qpois(0.9995,lambda)
```

```
> ylo
[1] 2
> yhi
[1] 22
> y1values <- ylo:yhi
> prob1 <- dpois(y1values,lambda)
> indic <- prob1 >= 0.001
> y1values <- y1values[indic]
> prob1 <- prob1[indic]
> sum(prob1)
[1] 0.9979123
> ylo
[1] 2
> yhi
[1] 22
> y1values
 [1]  2  3  4  5  6  7  8  9 10 11 12 13 14 15 16 17 18 19 20
>
```

As the sum of the probabilities that have been considered is 0.9979, which is very close to 1, I conclude that my choice of `qpois(0.0005,lambda)` and `qpois(0.9995,lambda)` for ylo and yhi, and of 0.001 for the lower bound for probabilities, is satisfactory. If `sum(prob1)` had been (say) 0.95 or less, I would have included more values of y in the sum by using `qpois(0.0001,lambda)` and `qpois(0.9999,lambda)` for ylo and yhi, and would have used 0.0001 for the lower bound for probabilities.

The procedure for determining an IMSE-optimal design in this situation is not dissimilar to that in Section 4.8. However, an important difference is the need to place lower and upper limits on the values of the explanatory variable x, as discussed in the first paragraph of Sub-section 2.4.2. Recall that, for a Bernoulli variable, $0 \leq \pi(\eta) \leq 1$. However, for a Poisson random variable, $0 < \lambda(\eta) < \infty$. A 10% discrepancy between $\pi^*(\eta)$ and $\pi(\eta)$ cannot exceed 0.1, whereas a 10% discrepancy between $\lambda^*(\eta)$ and $\lambda(\eta)$ can approach ∞ as $\lambda(\eta)$ approaches ∞.

Example 5.4.6. *Suppose that one wished to restrict the canonical variable $z = \beta_0 + \beta_1 x$ to the range $-2 \leq z \leq 8$, and therefore decided in evaluating the integral for IMSE(z) in* (5.24) *that $z_{lo} = -2$ and $z_{hi} = 8$. It was decided that the design would have two support points, z_1 and z_2, with $n_1 = n_2 = 4$. The MSE was calculated and plotted against z for several values of (z_1, z_2) to get a feel for what an appropriate initial guess might be for (z_1, z_2) when using* optim*. Figure 5.5 shows the plots for (z_1, z_2) equal to* (6.5, 8.0), (7.0, 8.0) *and* (7.5, 8.0)*. It can be seen that the area beneath the curve MSE(z) and above MSE(z) $= 0$ is least when*

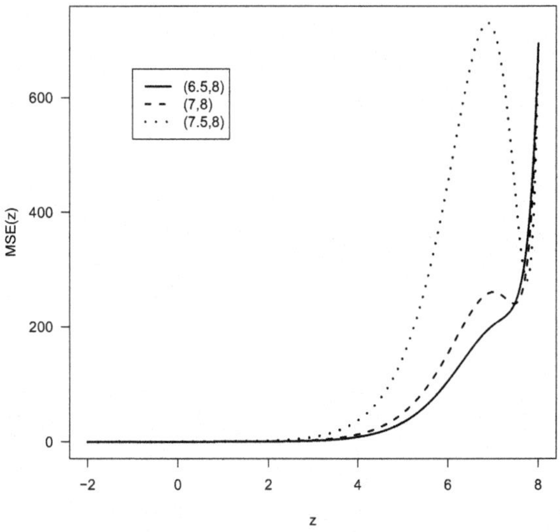

Figure 5.5 *Plots of MSE(*z*) vs* z *when* $n_1 = n_2 = 4$ *and the support points are at (i)* $(6.5, 8.0)$*, (ii)* $(7.0, 8.0)$ *and (iii)* $(7.5, 8.0)$.

the support points are 6.5 *and* 8.0*. This suggested that a search for the IMSE-optimal design could be given the starting points* 6.5 *and* 8.0.

A search was performed to find the IMSE-optimal design for the canonical variable z *on the domain* $-2 \leq z \leq 8$ *using limits of* `qpois(0.0001,lambda)` *and* `qpois(0.9999,lambda)` *for* ylo *and* yhi*, and 0.0001 for the lower bound for probabilities. To force* z *to lie between -2 and 8, values* w_1 *and* w_2 *were generated from the uniform distribution on* $(0, 1)$*, and then the transformation* $z = 3 + 5\cos(\pi w)$ *was used. The inverse of this transformation is* $w = \arccos[(z-3)/5]/\pi$*. Other than the restrictions on the values of* y_1 *and* y_2*, the program is similar to Program_14, which was described in Sub-section 4.8.3. The program to find the IMSE-optimal two-point design for the Poisson distribution is labelled Program_18, and appears in the online resources. As the slope of MSE(*z*) vs* z *is very steep near* $z = 8$*, the optimal design that you obtain and the minimum value of the MSE seem to depend on the value of the greater support point (which will frequently differ from 8 only in the fifth decimal place) found by* optim*. Notwithstanding this, the IMSE-optimal design for* $n_1 = n_2 = 4$ *is approximately* $z_1 = 6.271$ *and* $z_2 = 8$*, with an IMSE of 542.28.*

Let the values of $n_1, \dots, n_s$ be given. If $z_1, \dots, z_s$ form the support points of the IMSE-optimal design for $\eta = z$, the support points for the model $\eta = \beta_0 + \beta_1 x$ will be at

$$x_i = \frac{z_i - \beta_0}{\beta_1} \ (i = 1, \dots, s).$$

The minimum value of the IMSE, achieved for this design, will be $M/|\beta_1|$, where M is the value of the IMSE obtained for the design for the canonical variable.

Example 5.4.7. *For $\boldsymbol{\beta} = (1,2)^\top$, the IMSE-optimal design for $s = 2$ and $n_1 = n_2 = 4$ when $(-2-1)/2 < x < (8-1)/2$ (i.e., $-1.5 < x < 3.5$) has support points at $x_1 = (6.271 - 1)/2 = 2.636$ and $x_2 = (8-1)/2 =$* 3.5.

It is possible to extend the principle of IMSE-optimality for linear predictors $\eta = \beta_0 + \beta_1 x$ to $s > 2$ support points (as was done in the binomial situation in Section 4.8.2). However, the number of possible sampling results $(y_1, \dots, y_s)$ becomes very large as s increases (even with the most "pragmatic" approach) and makes the required computational time almost impractical.

Chapter 6

Several Other Distributions

6.1 Introduction

This chapter first describes how to apply a GLM to the multinomial distribution, even though the distribution does not belong to the exponential family of distributions. Then it considers how to find locally D-optimal designs for the multinomial distribution for various vectors of parameters.

After that, it provides a brief description of the considerations of designing an experiment when the response variable is thought to have a gamma distribution.

Finally, it considers situations where the nature of the distribution of the response variable is unknown, but it is believed that the link function and variance function can be specified. Analyses in this situation are often called quasi-likelihood analyses.

6.2 The multinomial distribution

6.2.1 Modelling data from a multinomial distribution

In a multinomial experiment, the response variable Y takes a value from one of a fixed number of categories. These categories may be nominal in nature (the categories are labels that cannot be put in a meaningful order), or ordinal (where there is a meaningful ordering). Examples of nominal categories are the political parties for which a person may vote. These parties may be listed in alphabetical order, but this is not a meaningful order with regard to (say) the parties' policies. If it were possible to order these parties from "most extreme left-wing views" to "most extreme right-wing views," this could be a meaningful ordering. Examples of ordinal categories are the severities of a disease suffered by patients attending an out-patient clinic: mild, moderate, and severe.

If there are k categories, they are frequently numbered from 1 to k, whether or not this ordering is meaningful.

Suppose that an experiment is carried out independently n times and that, on each occasion, the result will be exactly one of these k categories.

Let Y_i $(i = 1, \dots, k)$ be the number of times that an observation in category i is observed. The random variables $Y_1, \dots, Y_k$ are said to have a *multinomial distribution*

$$(Y_1, \dots, Y_k) \sim \text{Multinomial}(n; \pi_1, \dots, \pi_k)$$

where π_i $(i = 1, \dots, k)$ is the probability that an outcome is in category i. The observed values $y_1, \dots, y_k$ sum to n. The π_i satisfy

$$\pi_i > 0 \ (i = 1, \dots, k) \quad \text{and} \quad \sum_{i=1}^{k} \pi_i = 1.$$

The probability function for $(Y_1, \dots, Y_k)$ is

$$\begin{aligned} f(y_1, \dots, y_k;\ n, \pi_1, \dots, \pi_k) &= \Pr(Y_1 = y_1, Y_2 = y_2, \dots, Y_k = y_k) \\ &= \frac{n!}{y_1! y_2! \dots y_k!} \pi_1^{y_1} \pi_2^{y_2} \times \dots \times \pi_k^{y_k}. \end{aligned} \tag{6.1}$$

The binomial distribution is a multinomial distribution with $k = 2$.

For $k > 2$, (6.1) does not take the form in (1.13) required for it to be a member of the exponential family of distributions. However, it can be shown that GLMs can be used to model a multinomial situation. For example, see Dobson & Barnett (2008, Section 8.2) or Faraway (2006, Section 5.1).

The multinomial distribution differs from those distributions studied earlier in this book, as it has a multivariate random variable, and the individual components of $\boldsymbol{Y} = (Y_1, \dots, Y_k)^\top$ are not independent of one another. (While the numbers of successes and failures in a binomial distribution are not independent of one another, that complication is avoided by considering only the number of successes.) The components of $\boldsymbol{Y} = (Y_1, \dots, Y_k)^\top$ satisfy $\text{E}(Y_i) = n\pi_i$, $\text{var}(Y_i) = n\pi_i(1 - \pi_i)$ and $\text{cov}(Y_i, Y_j) = -n\pi_i\pi_j$ for $i, j \in \{1, \dots, k\}$ and $i \neq j$. The covariances are all negative because, for a fixed value of n, an increase in some Y_i means that another component Y_j $(j \neq i)$ must decrease.

As in previous chapters, it is assumed that there are m mathematically independent explanatory variables $x_1, \dots, x_m$ that may influence the values of the probabilities $\pi_1, \dots, \pi_k$. We wish to model $\pi_1, \dots, \pi_k$ in terms of $x_1, \dots, x_m$. As $\pi_1, \dots, \pi_k$ add to 1, their values are not independent of one another, so we do not use k linear predictors to model their individual values. It is customary to regard one of the probabilities (usually π_1 or π_k) as a *baseline probability,* and to express the remaining $(k - 1)$ probabilities in terms of the baseline probability using linear predictors.

For notational convenience, π_k will be treated as the baseline probability in this book.

Write

$$\eta_j = \ln\left(\frac{\pi_j}{\pi_k}\right) \quad (j = 1, \ldots, k-1) \tag{6.2}$$

for the jth linear predictor. Note that this is just a simple extension of the binomial model with a logit link: in that situation of two categories, if categories 1 and 2 represent "success" and "failure," respectively, then

$$\eta_1 = \ln\left(\frac{\pi_1}{\pi_2}\right) = \ln\left(\frac{\pi_1}{1-\pi_1}\right)$$

is just the standard logit link that was used in Chapter 4.

It is convenient to extend (6.2) to the case $j = k$. It follows that $\eta_k = \ln(1) = 0$, and we may write

$$\eta_j = \ln\left(\frac{\pi_j}{\pi_k}\right) \quad (j = 1, \ldots, k) \tag{6.3}$$

As in previous chapters, each η_j will be written as a linear combination of parameters, $\boldsymbol{f}_j^\top(\boldsymbol{x})\boldsymbol{\beta}$.

From (6.3),

$$\exp(\eta_j) = \pi_j/\pi_k, \quad (j = 1, \ldots, k) \tag{6.4}$$

and so

$$\begin{aligned} 1 + \sum_{\ell=1}^{k-1} \exp(\eta_\ell) &= 1 + \pi_1/\pi_k + \cdots + \pi_{k-1}/\pi_k \\ &= (\pi_1 + \pi_2 + \cdots + \pi_k)/\pi_k \\ &= 1/\pi_k. \end{aligned} \tag{6.5}$$

As (6.4) gives $\pi_j = \exp(\eta_j) \times \pi_k$ and (6.5) gives an expression for π_k, it may be deduced that

$$\pi_j = \frac{\exp(\eta_j)}{1 + \sum_{\ell=1}^{k-1} \exp(\eta_\ell)} \quad (j = 1, \ldots, k). \tag{6.6}$$

Thus the k probabilites $\pi_1, \ldots, \pi_k$ may be expressed in terms of the $(k-1)$ linear predictors $\eta_1, \ldots, \eta_{k-1}$ and $\eta_k = 0$.

Denote by $\boldsymbol{x} = (x_1, \ldots, x_m)^\top$ the vector of explanatory variables. Following Zocchi & Atkinson (1999), the vector of parameters, $\boldsymbol{\beta}$, and each $\boldsymbol{f}_i(\boldsymbol{x})$ $(i = 1, \ldots, k-1)$ are augmented by the addition of a zero as their last element, making them $(p+1) \times 1$ vectors. In addition, define $\boldsymbol{f}_k(\boldsymbol{x})$ to be the $(p+1) \times 1$ vector $(0, \ldots, 0, 1)^\top$. Then the linear predictors $\eta_1, \ldots, \eta_k$ may be written as

$$\eta_i = \boldsymbol{f}_i^\top(\boldsymbol{x})\boldsymbol{\beta} \quad (i = 1, \ldots, k). \tag{6.7}$$

Example 6.2.1. *A very simple model has $k = 3$ categories and $m = 1$ explanatory variable x, and the two linear predictors are $\eta_1 = \beta_1 + \beta_2 x$ and $\eta_2 = \beta_3 + \beta_4 x$. Then $\boldsymbol{\beta} = (\beta_1, \beta_2, \beta_3, \beta_4, 0)^\top$, $\boldsymbol{f}_1(x) = (1, x, 0, 0, 0)^\top$, $\boldsymbol{f}_2(x) = (0, 0, 1, x, 0)^\top$ and $\boldsymbol{f}_3(x) = (0, 0, 0, 0, 1)^\top$.*

Example 6.2.2. *For $k = 3$ categories and $m = 1$ explanatory variable x, Zocchi & Atkinson (1999) considered the model $\eta_1 = \beta_0 + \beta_1 x + \beta_2 x^2$ and $\eta_2 = \beta_3 + \beta_4 x$. It follows that $\boldsymbol{\beta} = (\beta_0, \beta_1, \beta_2, \beta_3, \beta_4, 0)^\top$, $\boldsymbol{f}_1(x) = (1, x, x^2, 0, 0, 0)^\top$, $\boldsymbol{f}_2(x) = (0, 0, 0, 1, x, 0)^\top$, and $\boldsymbol{f}_3(x) = (0, 0, 0, 0, 0, 1)^\top$.*

Example 6.2.3. *A simple model when there are $k = 3$ categories and $m = 2$ explanatory variables x_1 and x_2 has the linear predictors $\eta_1 = \beta_1 + \beta_2 x_1 + \beta_3 x_2$ and $\eta_2 = \beta_4 + \beta_5 x_1 + \beta_6 x_2$. Then $\boldsymbol{\beta} = (\beta_1, \beta_2, \beta_3, \beta_4, \beta_5, \beta_6, 0)^\top$, $\boldsymbol{f}_1(\boldsymbol{x}) = (1, x_1, x_2, 0, 0, 0, 0)^\top$, $\boldsymbol{f}_2(x) = (0, 0, 0, 1, x_1, x_2, 0)^\top$ and $\boldsymbol{f}_3(x) = (0, 0, 0, 0, 0, 0, 1)^\top$.*

Let $\boldsymbol{\eta}$ denote the $k \times 1$ vector and $\boldsymbol{F}$ the $k \times p$ matrix given respectively by

$$\boldsymbol{\eta} = \begin{bmatrix} \eta_1 \\ \eta_2 \\ \vdots \\ \eta_k \end{bmatrix} \quad \text{and} \quad \boldsymbol{F} = \begin{bmatrix} \boldsymbol{f}_1^\top(\boldsymbol{x}) \\ \boldsymbol{f}_2^\top(\boldsymbol{x}) \\ \vdots \\ \boldsymbol{f}_k^\top(\boldsymbol{x}) \end{bmatrix}.$$

Then (6.7) may be written as

$$\boldsymbol{\eta} = \boldsymbol{F}\boldsymbol{\beta}. \tag{6.8}$$

6.2.2 *Estimating the parameter vector $\boldsymbol{\beta}$*

For the multinomial distribution, the existence of several link functions and the relationship in (6.6) between π_j and the various η_i make estimation of $\boldsymbol{\beta}$ more complicated than it is for the binomial distribution. Consequently, it is more difficult to derive the equations that the ML estimate, $\hat{\boldsymbol{\beta}}$, must satisfy, and to obtain an expression for the information matrix $\boldsymbol{M}(\xi, \boldsymbol{\beta})$, than it was for the binomial distribution. It is easier to use formulae obtained by Zocchi & Atkinson (1999), on which the following development is based.

Equation (6.3) may be written in matrix notation as

$$\boldsymbol{\eta} = \boldsymbol{C}^\top \ln(\boldsymbol{L}\boldsymbol{\pi}), \tag{6.9}$$

where $\boldsymbol{L}$ is a matrix that selects the appropriate elements of $\boldsymbol{\pi}$ and the matrix $\boldsymbol{C}^\top$ selects appropriate multiples of each $\ln(\cdot)$. This notation is best explained through an example. Let $k = 3$. Then, noting that $\ln(a/b) = \ln(a) - \ln(b)$ and that $\ln(\pi_1 + \pi_2 + \pi_3) = \ln(1) = 0$, (6.3) says that

$$\begin{bmatrix} \eta_1 \\ \eta_2 \\ \eta_3 \end{bmatrix} = \begin{bmatrix} \ln(\pi_1) - \ln(\pi_3) \\ \ln(\pi_2) - \ln(\pi_3) \\ 0 \end{bmatrix}$$
$$= \begin{bmatrix} \ln(\pi_1) - \ln(\pi_3) \\ \ln(\pi_2) - \ln(\pi_3) \\ \ln(\pi_1 + \pi_2 + \pi_3) \end{bmatrix}$$
$$= \begin{bmatrix} 1 & 0 & -1 & 0 \\ 0 & 1 & -1 & 0 \\ 0 & 0 & 0 & 1 \end{bmatrix} \ln \left\{ \begin{bmatrix} 1 & 0 & 0 \\ 0 & 1 & 0 \\ 0 & 0 & 1 \\ 1 & 1 & 1 \end{bmatrix} \begin{bmatrix} \pi_1 \\ \pi_2 \\ \pi_3 \end{bmatrix} \right\};$$

that is, $\boldsymbol{\eta} = \boldsymbol{C}^\top \ln(\boldsymbol{L}\boldsymbol{\pi})$ as in (6.9).

Consider the standard notation for an approximate design:

$$\xi = \left\{ \begin{array}{cccc} \boldsymbol{x}_1 & \boldsymbol{x}_2 & \dots & \boldsymbol{x}_s \\ \delta_1 & \delta_2 & \dots & \delta_s \end{array} \right\}. \tag{6.10}$$

For a given value of $\boldsymbol{\beta}$, denote by $\boldsymbol{\pi}(\boldsymbol{x_i})$ the vector of probabilities $(\pi_1(\boldsymbol{x_i}), \dots, \pi_k(\boldsymbol{x_i}))^\top$ calculated from (6.7) and (6.6) at the ith support point, $\boldsymbol{x}_i$ $(i = 1, \dots, s)$.

From Zocchi & Atkinson (1999, p. 439), the $k \times k$ matrix whose (i, j) element is $(\partial\eta_i)/(\partial\pi_j)$ $(i, j = 1, \dots, k)$ is found from (6.9) to be

$$\frac{\partial\boldsymbol{\eta}}{\partial\boldsymbol{\pi}} = \boldsymbol{C}^\top \boldsymbol{D}^{-1} \boldsymbol{L}, \tag{6.11}$$

where

$$\boldsymbol{D} = \text{diag}(\boldsymbol{L}\boldsymbol{\pi}).$$

It is clear from (6.8) that

$$\frac{\partial\boldsymbol{\eta}}{\partial\boldsymbol{\beta}} = \boldsymbol{F}. \tag{6.12}$$

Consequently, for $\boldsymbol{\pi}(\boldsymbol{x}_i)$, it follows from (6.12) and (6.11) that the matrix $\boldsymbol{G}(\boldsymbol{x}_i)$, whose (i, j) element is $(\partial\pi_i)/(\partial\beta_j)$, is given by

$$\boldsymbol{G}(\boldsymbol{x}_i) = \left(\frac{\partial\boldsymbol{\eta}}{\partial\boldsymbol{\pi}}\right)^{-1} \left(\frac{\partial\boldsymbol{\eta}}{\partial\boldsymbol{\beta}}\right) = \left(\boldsymbol{C}^\top \boldsymbol{D}^{-1} \boldsymbol{L}\right)^{-1} \boldsymbol{F}. \tag{6.13}$$

Consider an exact design for which n_i observations are taken at the ith support point ($n_1 + \cdots + n_s = N$). Let $\boldsymbol{y}_i$ denote the $k \times 1$ vector whose jth element shows how many of the n_i observations taken at the ith support point fall into the jth category ($j = 1, \ldots, k$). Then the ML estimate $\hat{\boldsymbol{\beta}}$ satisfies the equation

$$\sum_{i=1}^{s} \boldsymbol{G}^\top(\boldsymbol{x}_i)\text{diag}[\pi_1^{-1}(\boldsymbol{x}_i), \ldots, \pi_k^{-1}(\boldsymbol{x}_i)]\boldsymbol{y}_i \Bigg|_{\boldsymbol{\beta}=\hat{\boldsymbol{\beta}}} = \boldsymbol{0}. \quad (6.14)$$

If you wish to use R to fit the model to some multinomial data, see Faraway (2006, Chapter 6).

The information matrix for the design in (6.10) and the parameter vector $\boldsymbol{\beta}$ is given by

$$\boldsymbol{M}(\xi, \boldsymbol{\beta}) = \sum_{i=1}^{s} \boldsymbol{G}^\top(\boldsymbol{x}_i)\text{diag}[\pi_1^{-1}(\boldsymbol{x}_i), \ldots, \pi_k^{-1}(\boldsymbol{x}_i)]\boldsymbol{G}(\boldsymbol{x}_i). \quad (6.15)$$

As always, the locally D-optimal design is that design ξ in the design space for which $\det[\boldsymbol{M}(\xi, \boldsymbol{\beta})]$ is maximised.

For an approximate design ξ, the standardised variance required for the general equivalence theorem is given (Zocchi & Atkinson, 1999, Appendix) by

$$d(\boldsymbol{x}, \xi, \boldsymbol{\beta}) = \text{tr}\left[\boldsymbol{\mathcal{V}}^{-1}(\boldsymbol{x})\boldsymbol{\mathcal{G}}^\top(\boldsymbol{x})\boldsymbol{\mathcal{M}}^{-1}(\xi, \boldsymbol{\beta})\boldsymbol{\mathcal{G}}^\top(\boldsymbol{x})\right], \quad (6.16)$$

where $\boldsymbol{\mathcal{V}}(\boldsymbol{x})$, $\boldsymbol{\mathcal{G}}(\boldsymbol{x})$, and $\boldsymbol{\mathcal{M}}(\xi, \boldsymbol{\beta})$ represent, respectively, the matrices $\boldsymbol{V}(\boldsymbol{x})$, $\boldsymbol{G}(\boldsymbol{x})$ and $\boldsymbol{M}(\xi, \boldsymbol{\beta})$ after their last row and column have been removed.

A design ξ^* is locally D-optimal if, for a given value of $\boldsymbol{\beta}$, the maximum value of $d(\boldsymbol{x}, \xi^*, \boldsymbol{\beta})$ is p for any $\boldsymbol{x}$ in the design region $\mathcal{X}$, and $d(\boldsymbol{x}, \xi^*, \boldsymbol{\beta})$ equals p at each support point $\boldsymbol{x}_i$ of ξ^*.

6.2.3 Some designs

Example 6.2.4. *First consider one explanatory variable. For $k = 3$ categories, and $m = 1$ and $s = 3$, consider the parameter vector $\boldsymbol{\beta} = (0, 1, 0, 1, 0)^\top$. This gives the model $\eta_1 = \eta_2 = 0 + 1 \times x = x$, which implies that $\pi_1 = \pi_2$ for each value of x in the design space, $\mathcal{X}$. Consider the standard situation where $\mathcal{X} = [-1, 1]$. The program below may be used to search for a locally D-optimal design.*

Segment ① specifies the number of categories (k), defines the matrices

$\boldsymbol{C}^\top$ and $\boldsymbol{L}$ introduced in (6.9), and defines the matrix $\boldsymbol{F}$ introduced immediately above (6.8). It also specifies the number of parameters, p, and creates the function that will calculate the negative of the determinant of the information matrix for a specified design.

①

```
betavec <- c(0,1,0,1,0)
k <- 3
ctrans <- matrix(c(1,0,-1,0,0,1,-1,0,0,0,0,1),3,4,byrow=T)
lmat <- matrix(c(1,0,0,0,1,0,0,0,1,1,1,1),4,3,byrow=T)
fmat <- function(x)
{
 mat <- matrix(c(1,x,0,0,0,0,0,1,x,0,0,0,0,0,1),3,5,byrow=T)
 mat
}
 p <- 4

detinfomat <- function(variables)
{
 infomat <- matrix(0,(p+1),(p+1))
 xvals <- cos(pi*variables[1:lim1])
 deltavec <- (variables[(lim1+1):lim2])^2
 deltavec <- deltavec/sum(deltavec)
 for (i in 1:s)
 {
  fmatx <- fmat(xvals[i])
  etavec <- as.vector(fmatx%*%betavec)
  temp <- exp(etavec)
  pivec <- temp/sum(temp)
  temp2 <- as.vector(lmat%*%pivec)
  dmatinv <- diag(1/temp2)
  temp3 <- ctrans%*%dmatinv%*%lmat
  temp3inv <- solve(temp3)
  gmat <- temp3inv%*%fmatx
  infomat <- infomat + deltavec[i]*t(gmat)%*%diag(1/pivec)%*%gmat
 }
 -det(infomat)
}
```

Segment ② of the program defines the values of m and s, specifies the number of starting values to be generated for searches of optimal designs, then generates initial values of z that can be transformed into values of the support points and design weights using the techniques described in Sub-section 2.4.3. The design ξ with the minimum value of $-det[\boldsymbol{M}(\xi, \boldsymbol{\beta})]$ (corresponding to the design with the maximum value of $det[\boldsymbol{M}(\xi, \boldsymbol{\beta})]$) is stored and printed out.

②

```
 m <- 1
 s <- 3
 lim1 <- m*s
 lim2 <- (m+1)*s
#simulations of different initial values
 nsims <- 200
 mindet <- 10
 #set.seed(123)
 for (i in 1:nsims)
 {
  initial <- runif(2*s)
  out <- optim(initial,detinfomat,NULL,method="Nelder-Mead")
  valuenow <- out$val
  if(valuenow < mindet) {mindet <- valuenow
  design <- out$par}
 }
 cat("Min value of det\n",mindet,"\n")
 output <- design
 out1 <- cos(pi*output[1:lim1])
 out2 <- (output[(lim1+1):lim2])^2
 wts <- out2/sum(out2)
 out4 <- cbind(matrix(out1,s,1),wts)
 out4 <- out4[order(out4[,1],out4[,2]),]
 cat("Design\n")
 t(out4)
```

If Segments ① and ② of the program are run as listed above, the result suggests that the locally D-optimal design is

$$\xi_1 = \left\{ \begin{array}{cc} -1.000 & 1.000 \\ 0.500 & 0.500 \end{array} \right\}.$$

The following program calculates the standardised variance $d(x, \xi_1, \boldsymbol{\beta})$ for this design. It assumes that the preceding program has already been run, so that the values of $\boldsymbol{\beta}$, p, $\boldsymbol{C}^\top$ etc. are already in the workspace. The program begins by constructing the information matrix for ξ_1. Then it defines a function that calculates $d(x, \xi_1, \boldsymbol{\beta})$. Then $d(x, \xi_1, \boldsymbol{\beta})$ is evaluated at each support point of ξ_1, and finally $d(x, \xi_1, \boldsymbol{\beta})$ is plotted against x for $x \in [-1, 1]$. The last four commands draw dotted lines and mark the support points.

```
s <- 2
info <- matrix(0,(p+1),(p+1))
xvals <- c(-1,1)
deltavec <- c(0.5,0.5)
for (i in 1:s)
{
```

```
 fmatx <- fmat(xvals[i])
 etavec <- as.vector(fmatx%*%betavec)
 temp <- exp(etavec)
 pivec <- temp/sum(temp)
 temp2 <- as.vector(lmat%*%pivec)
 dmatinv <- diag(1/temp2)
 temp3 <- ctrans%*%dmatinv%*%lmat
 temp3inv <- solve(temp3)
 gmat <- temp3inv%*%fmatx
 info <- info + deltavec[i]*t(gmat)%*%diag(1/pivec)%*%gmat
}
-det(info)
infosmall <- info[1:p,1:p]
invinfosmall <- solve(infosmall)

stdvar <- function(x)
{
 fmatx <- fmat(x)
 etavec <- as.vector(fmatx%*%betavec)
 temp <- exp(etavec)
 pivec <- temp/sum(temp)
 vmat <- diag(pivec) - pivec%*%t(pivec)
 temp2 <- as.vector(lmat%*%pivec)
 dmatinv <- diag(1/temp2)
 temp3 <- ctrans%*%dmatinv%*%lmat
 temp3inv <- solve(temp3)
 gmat <- temp3inv%*%fmatx
 vmatsmall <- vmat[1:(k-1),1:(k-1)]
 invvmatsmall <- solve(vmatsmall)
 gmatsmall <- gmat[1:(k-1),1:p]
 product <- invvmatsmall%*%gmatsmall%*%invinfosmall%*%t(gmatsmall)
 product
 sum(diag(product))
}
stdvar(-1)
stdvar(1)

#Plot the standardised variance
xvec <- seq(from=-1,to=1,by=0.01)#seq(from=-3,to=3,by=0.01)
yvec <- sapply(xvec,stdvar)
par(las=1)
plot(xvec,yvec,type="l",xlab="x",ylab="Standardised variance",lwd=2)
lines(c(-1.02,1.02),c(4,4),lty=2,lwd=2)
lines(c(-1,-1),c(2.45,4),lty=2,lwd=2)
lines(c(1,1),c(2.45,4),lty=2,lwd=2)
points(c(-1,1),c(4,4),pch=16,cex=2)
```

It is found that $d(-1, \xi_1, \boldsymbol{\beta}) = d(1, \xi_1, \boldsymbol{\beta}) = 4 = p$*. A plot of the standard-*

ised variance appears in Figure 6.1. It is evident that, over the domain $[-1, 1]$*,* $d(x, \xi_1, \boldsymbol{\beta})$ *achieves its maximum value of* $p = 4$ *at the two support points of* ξ_1*. Consequently, by the general equivalence theorem, the design* ξ_1 *is locally D-optimal.*

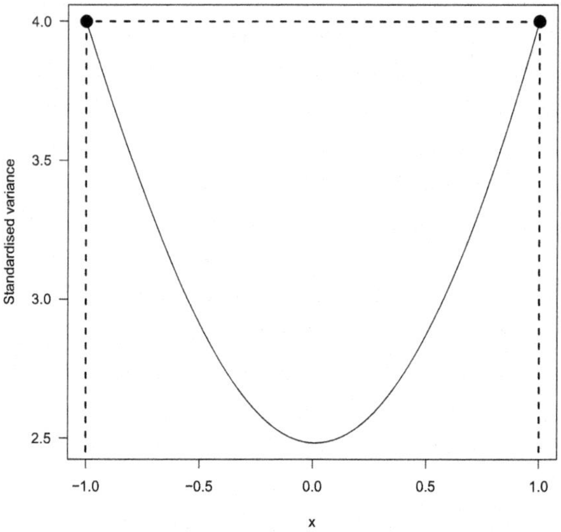

Figure 6.1 *Plot of* $d(x, \xi_1, \boldsymbol{\beta})$ *vs.* x *for* $x \in [-1, 1]$*. The maximum value of* $d(x, \xi_1, \boldsymbol{\beta})$ *is* $p = 4$ *and is achieved at the two support points (-1 and 1) of* ξ_1*.*

If the parameter vector is changed to $(0, 3, 0, 3, 0)^\top$, minor modifications to the preceding program will show that the locally D-optimal design is

$$\xi_2 = \left\{ \begin{array}{ccc} -0.562 & 0.183 & 1.000 \\ 0.418 & 0.217 & 0.365 \end{array} \right\}.$$

That is, it has three support points where the earlier design had two. A plot of the standardised variance appears in Figure 6.2.

Important note

Designs ξ_1 and ξ_2 both have less support points (two and three, respectively) than the value of p (four). This demonstrates a point made in the Comments on page 69. In both cases considered here, we have $\eta_1 = \eta_2$, so the information obtained from observations on category 1 contributes to estimating exactly the same quantity as the information from category

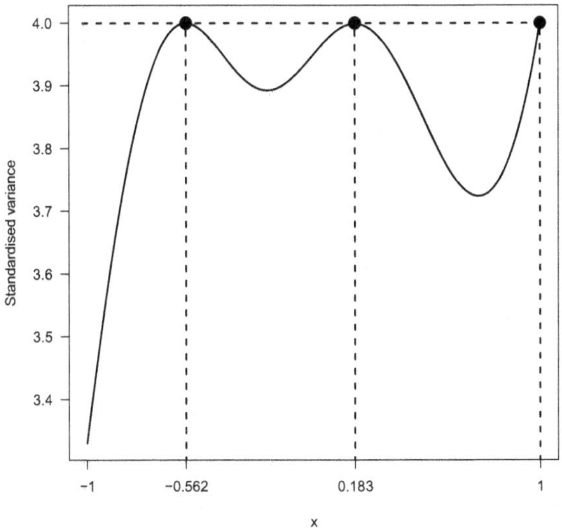

Figure 6.2 *Plot of $d(x, \xi_2, \boldsymbol{\beta})$ vs. x for $x \in [-1, 1]$. The maximum value of $d(x, \xi_2, \boldsymbol{\beta})$ is $p = 4$ and is achieved at the three support points of ξ_2.*

2. With additional information available to estimate the parameters, it should not be surprising that fewer support points are required.

We now consider $m = 2$ explanatory variables.

Example 6.2.5. *Let the parameter vector be $\boldsymbol{\beta} = (0, 1, 1, 1, 1, -1, 0)^\top$; i.e., $\eta_1 = x_1 + x_2$ and $\eta_2 = 1 + x_1 - x_2$. Only minor modifications are required to the R program beginning on page 177 in order to search for a locally D-optimal design.*

The following design, ξ_3, seems to be locally D-optimal. A contour plot of the standardised variance, $d(x, \xi_3, \boldsymbol{\beta})$, appears in Figure 6.3. The support points of ξ_3 have standardised variances of $p = 6$, and the standardised variance does not take a value greater than 6 in the design space. We may conclude that ξ_3 is the locally D-optimal design for the given model.

$$\xi_3 = \left\{ \begin{array}{cccc} (-1.000, -1.000)^\top & (-1.000, 1.000)^\top & (1.000, -0.135)^\top & (1.000, 1.000)^\top \\ 0.262 & 0.312 & 0.207 & 0.219 \end{array} \right\}.$$

Example 6.2.6. *Consider the design for the parameter set $\boldsymbol{\beta} = (1, 2, -1, -1, -1, 2, 0)^\top$. Computations using R suggest that the locally*

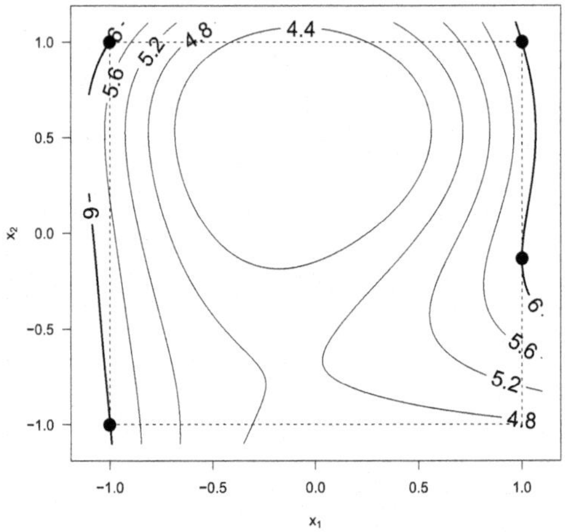

Figure 6.3 *Contour plot of* $d(x, \xi_3, \boldsymbol{\beta})$ *for* $\boldsymbol{x} \in [-1, 1]^2$. *The maximum value of* $d(x, \xi_3, \boldsymbol{\beta})$ *is* $p = 6$ *and is achieved at the support points of* ξ_3.

D-optimal design is

$$\xi_4 = \left\{ \begin{array}{rrrrrrr} x_1: & -1.000 & -1.000 & -1.000 & -0.192 & 0.005 & 0.764 \\ x_2: & -1.000 & -0.264 & 0.542 & -1.000 & 1.000 & 1.000 \\ \omega: & 0.259 & 0.083 & 0.232 & 0.059 & 0.067 & 0.300 \end{array} \right\}.$$

It was quite challenging to find the design ξ_4*, as the R program would repeatedly produce designs with less than six support points. However, the contour plot of the relevant standardised variance would show that there were other points in the design space with greater values of the standardised variance than the support points, meaning that the design could not be locally D-optimal. It cannot be stressed too many times that a design should not be concluded to be locally D-optimal until after the standardised variance has been examined.*

Figure 6.4 displays the contour plot of $d(x, \xi_4, \boldsymbol{\beta})$. *There are no values of the standardised variance greater than* $p = 6$, *and* $d(x, \xi_4, \boldsymbol{\beta}) = 6$ *at each of the support points. It may therefore be concluded that* ξ_4 *is the locally D-optimal design for the model under consideration.*

Further work on designs for the multinomial distribution, including a chapter on IMSE-optimality, may be found in Thompson (2010).

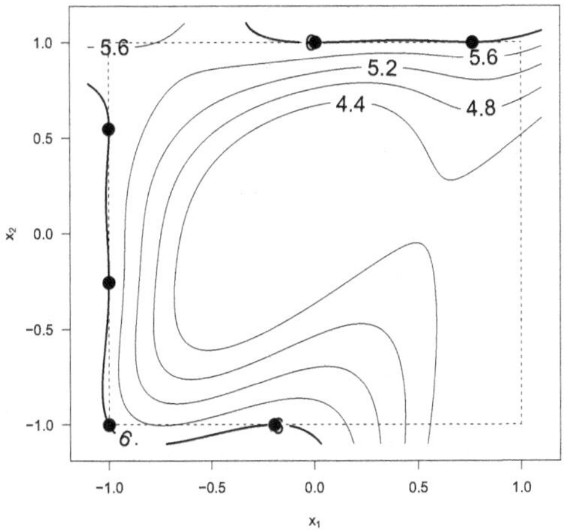

Figure 6.4 *Contour plot of* $d(x, \xi_4, \boldsymbol{\beta})$ *for* $\boldsymbol{x} \in [-1,1]^2$. *The maximum value of* $d(x, \xi_4, \boldsymbol{\beta})$ *is* $p = 6$ *and is achieved at the support points of* ξ_4.

6.3 The gamma distribution

The gamma distribution was introduced in Example 1.4.3, where it was shown to be a member of the exponential family of distributions. Unlike the Bernoulli, Poisson, and multinomial distributions, it is a continuous distribution. It has applications in areas such as insurance and queueing models, as it can model the time between events (contiguous or otherwise) that occur in accordance with a Poisson process.

The form of the probability function given in (1.14) is not always the most convenient for the purposes of a GLM (Faraway, 2006, p. 135). On page 14, it was stated that $\mu = \alpha/\beta$ and $\text{Var}(Y) = (1/\alpha)\mu^2$. As $\mu = \alpha/\beta$ implies that $\beta = \alpha/\mu$, β can be replaced by α/μ in (1.14) to give the alternative probability function

$$f_Y(y) = \frac{1}{\Gamma(\alpha)} \left(\frac{\alpha}{\mu}\right)^{\alpha} y^{\alpha-1} \exp(-\alpha y/\mu), \quad y > 0 \quad (\alpha > 0, \mu > 0). \tag{6.17}$$

The gamma distribution has the canonical form (page 14), so the canonical link is $g(\mu) = -\beta = -\alpha/\mu$ from Table 1.1. However, as $-\alpha$ is a known constant, it is acceptable to remove $-\alpha$ and to use the *reciprocal* function $g(\mu) = 1/\mu$ as the link function. This is the default link function used by the function `glm` in R when the *gamma* family is specified. This link function is not entirely satisfactory. The model requires that $\mu > 0$, but $\eta = g(\mu) = 1/\mu$ implies that $\mu = 1/\eta$, which might be negative. An alternative link function is the logarithmic function $g(\mu) = \ln(\mu)$, and $\eta = g(\mu)$ implies that $\mu = \exp(\eta)$, which is always nonnegative.

Consider the model weights $\omega_r(\boldsymbol{x}_i)$ and $\omega_\ell(\boldsymbol{x}_i)$ at a point $\boldsymbol{x}_i$ for the reciprocal and logarithmic links, respectively. Note that $\eta = g(\mu) = 1/\mu$ implies that $(\partial\eta)/(\partial\mu) = -1/\mu^2$ or $(\partial\mu)/(\partial\eta) = -\mu^2$. Hence, for the reciprocal link function,

$$\omega_r(\boldsymbol{x}_i) = \frac{1}{\text{Var}(Y_i)}\left(\frac{\partial\mu_i}{\partial\eta_i}\right)^2 = \frac{a}{\mu^2}\mu^4 = \alpha\mu^2.$$

The link function $\eta = g(\mu) = \ln(\mu)$ implies that $(\partial\eta)/(\partial\mu) = 1/\mu$ or $(\partial\mu)/(\partial\eta) = \mu$. With the log link function, the model weight is

$$\omega_\ell(\boldsymbol{x}_i) = \frac{1}{\text{Var}(Y_i)}\left(\frac{\partial\mu_i}{\partial\eta_i}\right)^2 = \frac{a}{\mu^2}\mu^2 = \alpha.$$

Note that $\omega_\ell(\boldsymbol{x}_i)$ does not depend on $\boldsymbol{\beta}$ at all, so that *any* D-optimal design for the gamma distribution found with the logarithmic link will be *globally optimal.*

When comparing the determinants of the $p \times p$ information matrices of competing designs, the term α in each of $\omega_r(\boldsymbol{x}_i)$ and $\omega_\ell(\boldsymbol{x}_i)$ may be ignored, as it is a constant common multiplier for all information matrices considered. This is equivalent to the approach for a normal distribution with the identity link, where $\omega(\boldsymbol{x}_i) = 1/\sigma^2$. D-optimal designs for the normal distribution are discussed in many books, including Atkinson, Donev, & Tobias (2007) and Box, Hunter, & Hunter (2005).

To find a locally D-optimal design using the *reciprocal* link, you may use the program described for the binomial distribution on pages 106 to 108 but with the following minor change. In the function *detinfomat2*, replace the two lines

```
expetavec <- exp(etavec)
modelwtvec <- expetavec/((1+expetavec)^2)
```

by the one line

```
modelwtvec <- 1/etavec^2
```

When calculating the standardised variance, the same changes must be made in the commands to determine the information matrix for the design thought to be locally D-optimal, and in the function *stdvar* which calculates the standardised variance at an individual value of $\boldsymbol{x}$.

To find a locally D-optimal design using the *logarithmic* link, use the same program on pages 106 to 108. However, in the function *detinfomat2*, replace the three lines

```
etavec <- as.vector(t(betavec)%*%fxmat)
expetavec <- exp(etavec)
modelwtvec <- expetavec/((1+expetavec)^2)
```

by the one line

```
modelwtvec <- rep(1,s)
```

As above, when calculating the standardised variance, the same changes must be made in the commands to determine the information matrix for the design thought to be locally D-optimal, and in the function *stdvar* which calculates the standardised variance at an individual value of $\boldsymbol{x}$.

Example 6.3.1. *Suppose that there are $m = 2$ predictor variables x_1 and x_2 that satisfy the usual constraints $-1 \leq x_i \leq 1$ $(i = 1, 2)$. We consider the linear predictor $\eta = 4 + 2x_1 + x_2$; i.e., the parameter vector is $\boldsymbol{\beta} = (4, 2, 1)^\top$. I sought D-optimal designs for both the reciprocal and log links. As there are $p = 3$ parameters, the number of support points for the optimal design will lie between $p = 3$ and $p(p+1)/2 = 6$. I started my search with $s = 6$ support points, and reduced the value of s if this was indicated. For each link, the search proceeded smoothly, and the following designs were found:*

reciprocal link:

$$\xi_r = \left\{ \begin{array}{ccc} (-1,-1)^\top & (-1,1)^\top & (1,-1)^\top \\ \frac{1}{3} & \frac{1}{3} & \frac{1}{3} \end{array} \right\}, \tag{6.18}$$

logarithmic link:

$$\xi_\ell = \left\{ \begin{array}{cccc} (-1,-1)^\top & (-1,1)^\top & (1,-1)^\top & (1,1)^\top \\ 0.25 & 0.25 & 0.25 & 0.25 \end{array} \right\}, \tag{6.19}$$

The contour plots of the standardised variances of ξ_r and ξ_ℓ appear in Figure 6.5. It is evident that the standardised variances achieve their maximum value of $p = 3$ at the support points of the relevant design, confirming that the designs are locally and globally D-optimal respectively.

Please note that the linear predictor $\eta = 4 + 2x_1 + x_2$ is always greater than 0 in the design region. This ensures that $\mu > 0$ for the reciprocal link everywhere in the region. However, should $\boldsymbol{\beta}$ be such that the line $\eta = 0$ passes through the design region, then the region will contain a sub-space where $\mu < 0$. As well, $\omega_r(\boldsymbol{x}_i) = 1/\eta^2$ will be infinite whenever $\eta = 0$, and this will cause the maximisation of det$[\boldsymbol{M}(\xi, \boldsymbol{\beta})]$ to yield

(a)

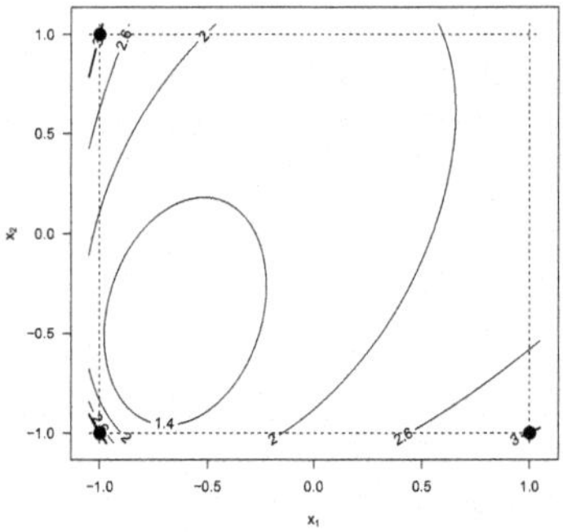

(b)

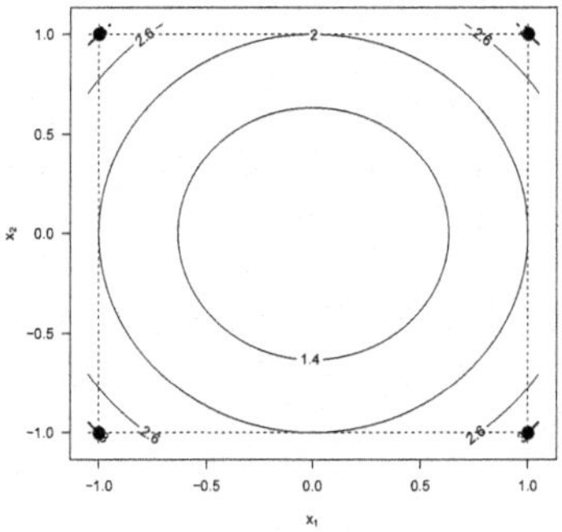

Figure 6.5 *Contour plot of (a)* $d(\boldsymbol{x}, \xi_r, \boldsymbol{\beta})$ *and (b)* $d(\boldsymbol{x}, \xi_\ell, \boldsymbol{\beta})$ *for* $\boldsymbol{x} \in [-1, 1]^2$*. The designs are given in* (6.18) *and* (6.19)*. The maximum values of* $d(\boldsymbol{x}, \xi_r, \boldsymbol{\beta})$ *and* $d(\boldsymbol{x}, \xi_\ell, \boldsymbol{\beta})$ *are* $p = 3$ *and are achieved at the support points of the relevant design.*

nonsensical answers. If there is any chance that the linear predictor will yield values of zero, it would be advisable to select another link function.

Atkinson, Donev, & Tobias (2007, p. 410) recommend the Box and Cox family

$$g(\mu) = \begin{cases} (\mu^\lambda - 1)/\lambda & (\lambda \neq 0) \\ \ln \mu & (\lambda = 0) \end{cases} \tag{6.20}$$

as a "useful, flexible family of links." The quantity λ is a constant to be selected. For all values of λ, the model weights for this family satisfy

$$\omega(\boldsymbol{x}_i) = \mu_i^{-2\lambda}.$$

If $\eta = g(\mu)$, then we have

$$\mu = \begin{cases} (\lambda\eta + 1)^{1/\lambda} & (\lambda \neq 0) \\ \exp(\eta) & (\lambda = 0). \end{cases} \tag{6.21}$$

Remember that we require $\mu > 0$ for the gamma distribution. Do not choose a nonzero value of λ that will cause $(\lambda\eta + 1)$ to be negative over the design region.

Finding a locally D-optimal design for a GLM with a gamma distribution and this link function requires minimal modifications to the program on pages 106 to 108. After assigning a value to λ (e.g., `lambda <- 2`), replace the lines

```
expetavec <- exp(etavec)
modelwtvec <- expetavec/((1+expetavec)^2)
```

by the three lines

```
if(lambda == 0) {muvec <- exp(etavec)} else
  {muvec <- (lambda*etavec + 1)^(1/lambda)}
modelwtvec <- muvec^(-2*lambda)
```

and remember to make the same change in the calculation of $\boldsymbol{M}(\xi, \boldsymbol{\beta})$ for the design, ξ, thought to be locally D-optimal, and in the function *stdvar* which calculates the standardised variance at an individual value of $\boldsymbol{x}$.

Example 6.3.2. *Example 6.3.1 considered the reciprocal and logarithmic links for use with a gamma distribution. These correspond to* $\lambda = -1$ *(roughly) and* $\lambda = 0$ *in the Box and Cox transformations. I decided to use* $\lambda = -0.5$ *as an intermediate value, which from* (6.21) *requires* $(-0.5\eta + 1)$ *to be nonnegative over the design range, or* $\eta \leq 2$*. So I chose* $\boldsymbol{\beta} = (-3, 2, 1)^{\top}$*, as* $\eta = -3 + 2x_1 + x_2$ *is not positive over the design region. I found the locally D-optimal design to be*

$$\xi_{BC} = \left\{ \begin{array}{ccc} (-1,1)^{\top} & (1,-1)^{\top} & (1,1)^{\top} \\ \frac{1}{3} & \frac{1}{3} & \frac{1}{3} \end{array} \right\}. \tag{6.22}$$

Note that it has three support points, like the design for the reciprocal link, but that the support points are not identical to those of the design in (6.18).

A plot of the standardised variance of ξ_{BC} *over the design region appears in Figure 6.6. This makes it clear that* ξ_{BC} *is indeed locally D-optimal for the particular link.*

Note that the Box and Cox link functions are not standard options for link functions in *glm* in R. If you decide to use a Box and Cox link function in a data analysis using *glm*, be sure to consult an expert R programmer in advance, so that a way of using this link function can be devised.

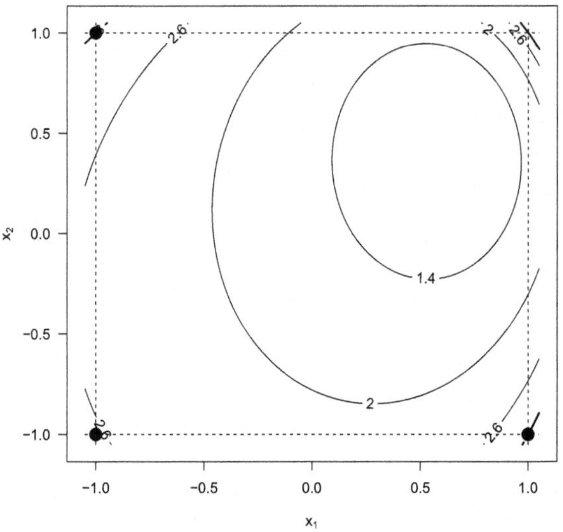

Figure 6.6 *Contour plot of* $d(\boldsymbol{x}, \xi_{BC}, \boldsymbol{\beta})$ *for* $\boldsymbol{x} \in [-1,1]^2$. *The design is given in* (6.22). *The maximum value of* $d(\boldsymbol{x}, \xi_{BC}, \boldsymbol{\beta})$ *is* $p = 3$ *and is achieved at the support points of the design.*

6.4 No specified distribution

There are occasions when the distribution of the response variable is unknown, but the experimenter believes that the link function $g(\cdot)$ and variance function $V(\mu)$ can be specified. Without knowledge of the distribution, the log likelihood function $\ell(\boldsymbol{\beta})$ cannot be constructed. However, a quasi-likelihood may be formulated. The mathematical details are provided in McCullagh & Nelder (1989, Chapter 9). To perform an analysis in R, one needs to make some modifications to the arguments of the *glm* function in R. An example is given in Faraway (2006, Section 7.4).

Faraway (2006, Section 3.1) contains the analysis of data from a dataset called *gala* (for Galapagos Islands) under the assumption that the response variable has a Poisson distribution. Running the commands

```
data(gala)
gala <- gala[,-2]
modp <- glm(Species~.,family = poisson,data = gala)
summary(modp)
```

gives the output

```
Deviance Residuals:
```

```
    Min       1Q   Median       3Q      Max
-8.2752  -4.4966  -0.9443   1.9168  10.1849

Coefficients:
              Estimate Std. Error z value Pr(>|z|)
(Intercept)  3.155e+00  5.175e-02  60.963  < 2e-16 ***
Area        -5.799e-04  2.627e-05 -22.074  < 2e-16 ***
Elevation    3.541e-03  8.741e-05  40.507  < 2e-16 ***
Nearest      8.826e-03  1.821e-03   4.846 1.26e-06 ***
Scruz       -5.709e-03  6.256e-04  -9.126  < 2e-16 ***
Adjacent    -6.630e-04  2.933e-05 -22.608  < 2e-16 ***
---
Signif. codes:  0 *** 0.001 ** 0.01 * 0.05 . 0.1   1

(Dispersion parameter for poisson family taken to be 1)

    Null deviance: 3510.73  on 29  degrees of freedom
Residual deviance:  716.85  on 24  degrees of freedom
AIC: 889.68

Number of Fisher Scoring iterations: 5
```

If, instead, the Poisson distribution is not specified, but it is still assumed that the link function is $g(\eta) = \ln(\eta)$ and the variance function is of the form $\text{var}(Y) = \phi V(\mu)$ for $V(\mu) = \mu$, any of the following three sets of commands

```
modq1 <- glm(Species~.,family = quasipoisson(link = "log"),gala)
summary(modq1)

modq2 <- glm(Species~.,family = quasipoisson,gala)
summary(modq2)

modr <- glm(Species~.,family = quasi(link = "log",
   variance = "mu"),gala)
summary(modr)
```

will give the output

```
Deviance Residuals:
    Min       1Q   Median       3Q      Max
-8.2752  -4.4966  -0.9443   1.9168  10.1849

Coefficients:
              Estimate Std. Error t value Pr(>|t|)
(Intercept)  3.1548079  0.2915901  10.819 1.03e-10 ***
Area        -0.0005799  0.0001480  -3.918 0.000649 ***
Elevation    0.0035406  0.0004925   7.189 1.98e-07 ***
Nearest      0.0088256  0.0102622   0.860 0.398292
Scruz       -0.0057094  0.0035251  -1.620 0.118380
```

```
Adjacent    -0.0006630  0.0001653  -4.012 0.000511 ***
---
Signif. codes:  0 *** 0.001 ** 0.01 * 0.05 . 0.1  1

(Dispersion parameter for quasipoisson family taken to be 31.74921)

    Null deviance: 3510.73  on 29  degrees of freedom
Residual deviance:  716.85  on 24  degrees of freedom
AIC: NA
```

(with the word "quasipoisson" replaced by "quasi" in *Dispersion parameter for* ... in the output from the fourth set of commands). The only numerical differences between the first table and subsequent tables occur in the values of *Pr*($>|t|$) and the dispersion parameter, ϕ, in $\text{var}(Y) = \phi V(\mu) = \phi\mu$. In the standard Poisson analysis, the value of ϕ is set equal to 1; in a quasi-likelihood analysis, the value of ϕ is estimated from the data.

The important thing to notice from the analyses is that only $g(\eta)$ and $V(\mu)$ are required. Given that the form of $\boldsymbol{f}(\boldsymbol{x})$ in $\eta = \boldsymbol{f}^\top(\boldsymbol{x})\,\boldsymbol{\beta}$ has been selected, then $g(\eta)$ and $V(\mu)$ are the only other pieces of information needed to calculate the information matrix $\boldsymbol{M}(\xi, \boldsymbol{\beta})$ in (3.6).

It follows that we need no additional procedures in finding a D- or D_S-optimal design for a quasi-likelihood situation. For example, a design that is locally D-optimal for a Poisson family will also be D-optimal for a design whose distribution is unknown but which has the same functions $g(\eta)$ and $V(\mu)$ as the Poisson distribution.

Chapter 7

Bayesian Experimental Design

7.1 Introduction

Bayesian analysis is a large and rapidly growing field of statistics. An oversimplified description is to say that, when estimating some parameter(s), one begins with an expression of *prior* knowledge or belief about the value of the parameters. This is combined with empirical knowledge about the value of the parameters that is gained by conducting an experiment. The result is a *posterior* description of the statistical behaviour of the parameters. Many books, including Carlin & Louis (2009), provide a coverage of Bayesian analysis.

Estimation of parameter values does not occur in the determination of an optimal design. However, the expression *Bayesian experimental design* is often used for the procedure about to be described because we use prior belief about the value of the vector of parameters, $\boldsymbol{\beta}$, to assist us in the determination of the design.

Of course, we have been expressing prior belief about the value of $\boldsymbol{\beta}$ in previous chapters, when using the nominated value of $\boldsymbol{\beta}$ as though we are certain that it is correct. In what follows, we recognise some uncertainty about our knowledge of $\boldsymbol{\beta}$. Many people who feel uncomfortable specifying the value of the parameter vector $\boldsymbol{\beta}$ in the methods considered in previous chapters may feel happier using Bayesian design methods.

An approach that is fully Bayesian, often referred to as a decision-theoretic approach, was reviewed by Chaloner & Verdinelli (1995), and is addressed by Overstall & Woods (2017). Consult either of these references for more information. The aim of this approach is to maximise the experimenter's gain from using one of a collection of designs if given one of a collection of vectors of response variables when the parameter vector is one from a collection of such vectors. It requires the maximisation of an expected utility function (Overstall & Woods, 2017, eq. 1)

$$U(\boldsymbol{\delta}) = \int\int_{\Psi,\mathcal{Y}} u(\boldsymbol{\delta}, \boldsymbol{\psi}, \boldsymbol{y})\pi(\boldsymbol{y}, \boldsymbol{\psi}|\boldsymbol{\delta})\, d\boldsymbol{y}\, d\boldsymbol{\psi}, \tag{7.1}$$

where the symbols are defined in Overstall & Woods (2017, p. 458).

You will not be surprised that Overstall & Woods (2017) say (p. 458) that "Selection of a fully Bayesian optimal design ... has traditionally been challenging for all but the most straightforward utility functions and models due to the high-dimensional and, typically, analytically intractable integrals" in (7.1).

To avoid the challenges of the "full Bayesian" decision-theoretic approach, we take a "pseudo-Bayesian" approach which is much less computationally demanding.

7.2 A discrete prior distribution for $\boldsymbol{\beta}$

7.2.1 *Finding the D-optimal design*

Suppose that h different values, $\boldsymbol{\beta}_1, \ldots, \boldsymbol{\beta}_h$, are believed to be possible values of the parameter vector $\boldsymbol{\beta}$. Associated with these values are h probabilities $\psi_1, \ldots, \psi_h$, respectively, satisfying $\psi_i > 0\,(i = 1, \ldots, h)$ and $\psi_1 + \cdots + \psi_h = 1$. These represent the strength of belief that each particular $\boldsymbol{\beta}_i$ is the true value of $\boldsymbol{\beta}$. These quantities can be written as a probability function

Parameter vector	$\boldsymbol{\beta}_1$	$\boldsymbol{\beta}_2$	...	$\boldsymbol{\beta}_h$
Prior probability	ψ_1	ψ_2	...	ψ_h

which is called the *prior probability distribution* of $\boldsymbol{\beta}$. The work in previous chapters, when locally D-optimal designs were calculated, represents the special case $h = 1$ and $\psi_1 = 1$ of Bayesian experimental design.

An important reference for Bayesian experimental design is Chaloner & Verdinelli (1995) which, in Section 4, considers nonlinear design problems. These include the designs for GLMs that we are considering. Chaloner & Larntz (1989) and Woods et al. (2006) have used Bayesian D-optimality to find designs for GLMs.

Adaptation of Chaloner & Verdinelli (1995, eq. 15) to the notation used here shows that a design ξ will be *Bayesian D-optimal* amongst a set of designs Ξ if it is the particular design that maximises the *utility function*

$$\phi(\xi) = \sum_{i=1}^{h} \psi_i \ln\{\det[\boldsymbol{M}(\xi, \boldsymbol{\beta}_i)]\}. \tag{7.2}$$

By adapting a result from Chaloner & Verdinelli (1995, p. 289), the standardised variance to be used in checking whether a candidate design ξ^* is indeed Bayesian D-optimal is

$$d(\boldsymbol{x}, \xi^*) = \sum_{i=1}^{h} \psi_i\, \omega(\boldsymbol{x}, \boldsymbol{\beta}_i) \boldsymbol{f}^\top(\boldsymbol{x}) \boldsymbol{M}^{-1}(\xi^*, \boldsymbol{\beta}_i) \boldsymbol{f}(\boldsymbol{x}). \tag{7.3}$$

Note that (7.2) and (7.3) both involve the calculation of h information matrices. This necessitates careful programming if the determination of a Bayesian D-optimal design and checking that it is indeed D-optimal (by use of the general equivalence theorem) are not to be unnecessarily repetitive.

Example 7.2.1. *Consider a single explanatory variable, x. For a logistic regression with $\eta = \beta_0 + \beta_1 x$ and $\boldsymbol{\beta} = (0,1)^\top$, it was given in (3.26) that the locally D-optimal design is*

$$\xi = \left\{ \begin{array}{cc} -1.5434 & 1.5434 \\ 0.5 & 0.5 \end{array} \right\}. \tag{7.4}$$

Suppose that the experimenter is uncertain about using $\boldsymbol{\beta} = (0,1)^\top$, and instead wishes to consider $h = 4$ alternatives $\boldsymbol{\beta}_1 = (-0.2, 0.8)^\top$, $\boldsymbol{\beta}_2 = (-0.2, 1.2)^\top$, $\boldsymbol{\beta}_3 = (0.2, 0.8)^\top$ and $\boldsymbol{\beta}_4 = (0.2, 1.2)^\top$, with equal prior probabilities; i.e., $\psi_1 = \ldots = \psi_4 = 0.25$. In the absence of other information, she uses $\mathcal{X} = \{x : -10 \leq x \leq 10\}$ as the design space.

The following program for $m = 1$ appears in the Web site doeforglm.com as Program_19:

```
f <- function(x)
{
 f <- c(1,x)
 f
}

infomat <- function(betavec,xvec,deswts)
{
 xmat <- t(matrix(xvec,s,m))
 fxmat <- apply(xmat,2,fx)
 etavec <- as.vector(t(betavec)%*%fxmat)
 expetavec <- exp(etavec)
 modelwtvec <- expetavec/((1+expetavec)^2)
 infomat <- fxmat %*% diag(deswts*modelwtvec) %*% t(fxmat)
 infomat
}

combine <- function(values)
{ #This function calculates the NEGATIVE of the sum of
  #log(determinant(infomat)) over all the potential beta vectors
 xvec <- 10*cos(pi*values[1:lim1])
 temp <- (values[(lim1+1):lim2])^2
 deltavec <- temp/sum(temp)
 sum <-  0
 for (m in 1:h)
 {
  betavec <- betamat[m,]
```

```
  logdetinfo <- log(det(infomat(betavec,xvec,deltavec)))
  sum <- sum - psivec[m]*logdetinfo
 }
 sum
}
```

The h values of $\boldsymbol{\beta}$ are entered as the rows of the $h \times p$ matrix betamat.

```
betamat <- matrix(c(-0.2,-0.2,0.2,0.2,0.8,1.2,0.8,1.2),4,2)
psivec <- rep(0.25,4)
h <- (dim(betamat))[1]
s <- 2
m <- 1
p <- 2
lim1 <- m*s
lim2 <- (m+1)*s

nsims <- 10000
min <- 100
for (isim in 1:nsims)
{
 values <- runif(lim2)
 out <- optim(values,combine,NULL,method="Nelder-Mead")
 temp <- out$val
 if (temp < min) {min <- temp
 design <- out$par}
}
points <- 10*cos(pi*(design[1:lim1]))
weights <- (design[(lim1+1):lim2])^2
weights <- weights/sum(weights)
display <- rbind(points,weights)
display
cat("min value of (-logdeterminant)",min,"\n")
```

The program's output suggests that the following design is D-optimal:

$$\xi_1 = \left\{ \begin{array}{cc} -1.5356 & 1.5357 \\ 0.5 & 0.5 \end{array} \right\}.$$

The changes from the earlier design in (7.4) to this one are minimal, which is not surprising, given that $\boldsymbol{\beta}_1, \ldots, \boldsymbol{\beta}_4$ surround $\boldsymbol{\beta} = (0,1)^\top$.

If the prior probabilities of $\boldsymbol{\beta}_1, \ldots, \boldsymbol{\beta}_4$ are changed from $\psi_1 = \ldots = \psi_4 = 0.25$ to $\psi_1 = 0.1$, $\psi_2 = 0.2$, $\psi_3 = 0.3$ and $\psi_4 = 0.4$, the program suggests that the locally D-optimal Bayesian design is

$$\xi_2 = \left\{ \begin{array}{cc} -1.5529 & 1.4004 \\ 0.5 & 0.5 \end{array} \right\}.$$

The following program is Program_20 in the online resources, and can be easily generalised. It will calculate the standardised variance $d(x, \xi_2)$ at the support points $x_1 = -1.5529$ and $x_2 = 1.4004$, and then plot $d(x, \xi_2)$ vs. x for all values of x between -5 and 5:

```
xvec <- c(-1.5529,1.4004)
deltavec <- c(0.5,0.5)
invinfomatrices <- rep(0,p*p*h)
dim(invinfomatrices) <- c(p,p,h)
for (m in 1:h)
{
 betavec <- betamat[m,]
 invinfomatrices[,,m] <- solve(infomat(betavec,xvec,deltavec))
}
invinfomatrices

stdvar <- function(x)
{
 sv <- 0
 fx <- f(x)
 for (m in 1:h)
 {
  betavec <- betamat[m,]
  eta <- sum(f(x)*betavec)
  expeta <- exp(eta)
  wt <- expeta/((1+expeta)^2)
  matinv <- invinfomatrices[,,m]
  sv <- sv + psivec[m]*wt*t(fx)%*%matinv%*%fx
 }
sv
}

stdvar(-1.5529)
stdvar(1.4004)
xvec <- seq(from=-5,to=5,by=0.01)
yvec <- sapply(xvec,stdvar)
plot(xvec,yvec,ty="l",xlab="x",ylab="Standardised variance")
lines(c(-5,5),c(2,2),lty=2)
```

It is assumed that this program runs in R immediately after the program on page 193; i.e., it uses all the functions defined, and output produced, in that program. In order to avoid calculating $\boldsymbol{M}^{-1}(\xi, \boldsymbol{\beta}_1), \ldots, \boldsymbol{M}^{-1}(\xi, \boldsymbol{\beta}_h)$ afresh for every value of x, I have created a $p \times p \times h$ array invinfomatrices, in which `invinfomatrices[,,j]` *contains the $p \times p$ matrix $\boldsymbol{M}^{-1}(\xi, \boldsymbol{\beta}_j)$. Each of the h matrices $\boldsymbol{M}^{-1}(\xi, \boldsymbol{\beta}_i)$ is calculated only once, no matter for how many values of x the value of $d(x, \xi)$ is calculated.*

The values of $d(-1.5529, \xi_2)$ and $d(1.4004, \xi_2)$ are each 2. A plot of

$d(x, \xi_2)$ vs. x for $x \in \{x : -5 \leq x \leq 5\}$ appears in Figure 7.1. It shows only two peaks, with values of two. This supports the belief that $d(x, \xi_2)$ achieves its maximum value of $p = 2$ at the two support points of ξ_2, and that therefore ξ_2 is the Bayesian D-optimal experimental design for the specified values of $\boldsymbol{\beta}_i$ and their prior probabilities ψ_i, $(i = 1, \ldots, 4)$.

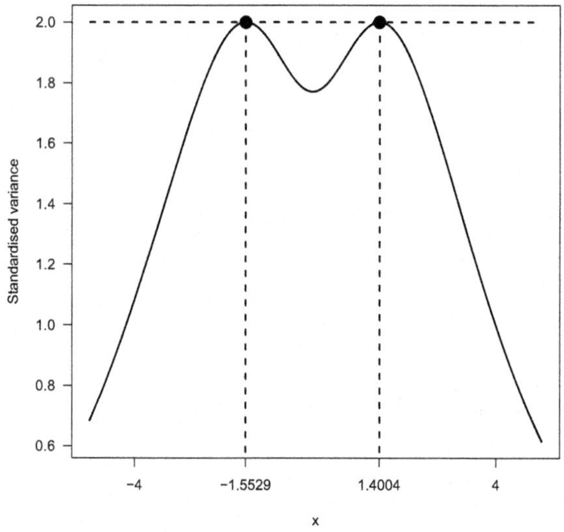

Figure 7.1 *Plot of the standardised variance $d(x, \xi_2)$ vs. x for $-5 \leq x \leq 5$.*

Example 7.2.2. *Consider a situation with $m = 2$ predictor variables, where the linear predictor is $\eta = \beta_0 + \beta_1 x_1 + \beta_2 x_2$, and both x_1 and x_2 are restricted to lie in the interval $\{x : -1 \leq x \leq 1\}$. Suppose that $h = 4$ values are specified for $\boldsymbol{\beta}$, namely $\boldsymbol{\beta}_1 = (-0.2, 0.8, 0.8)^\top$, $\boldsymbol{\beta}_2 = (-0.2, 1.2, 1.2)^\top$, $\boldsymbol{\beta}_3 = (0.2, 0.8, 1.2)^\top$ and $\boldsymbol{\beta}_4 = (0.2, 1.2, 0.8)^\top$, with prior probabilities $\psi_1 = 0.1$, $\psi_2 = 0.2$, $\psi_3 = 0.3$ and $\psi_4 = 0.4$, respectively.*

The program on page 193 is modified by changing the definition of the function `f`, the value of p, and the values of the matrix of $\boldsymbol{\beta}$ vectors and the vector of probabilities ψ_i. I initially started with $s = 8$ support points, but was able to reduce this to $s = 4$ in successive searches for the Bayesian D-optimal design. I eventually obtained the following design:

$$\xi_3 = \left\{ \begin{array}{cccc} (-1,-1)^\top & (-1,1)^\top & (0.9689,1)^\top & (1,-1)^\top \\ 0.2243 & 0.2958 & 0.1832 & 0.2967 \end{array} \right\}. \tag{7.5}$$

Note that I have not followed my usual practice of recording the values of the support points and design weights to three decimal places. This is because, when three decimal places were used, the plot of the standardised variance suggested that $d(\boldsymbol{x}, \xi_3)$ *was greater than 3 at* $\boldsymbol{x} = (1,1)^\top$*, which implies that* ξ_3 *is not D-optimal. However, when the matrices* $\boldsymbol{M}(\xi_3, \boldsymbol{\beta}_1), \ldots, \boldsymbol{M}(\xi_3, \boldsymbol{\beta}_4)$ *were calculated using the four decimal places given in* (7.5)*, the plot of the standardised variance was as shown in Figure 7.2. In this figure, it is clear that the maximum value of* $d(\boldsymbol{x}, \xi_3)$ *is* $p = 3$*, and occurs at the support points of the design. Hence it is concluded that* ξ_3 *is the Bayesian D-optimal experimental design.*

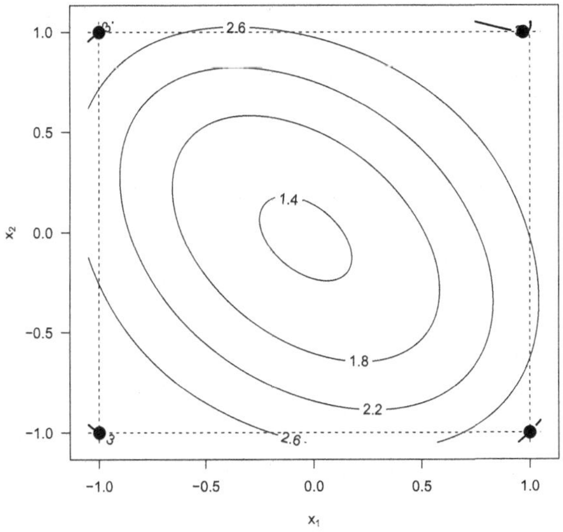

Figure 7.2 *Plot of the standardised variance* $d(\boldsymbol{x}, \xi_3)$ *for the Bayesian D-optimal design* ξ_3 *given in* (7.5)*. Within the design space,* $d(\boldsymbol{x}, \xi_3)$ *achieves the required maximum value of* $p = 3$ *at each of the support points.*

7.2.2 Next steps

Having found the Bayesian D-optimal design, the next step is to run an experiment and collect some data that will permit the estimation of $\boldsymbol{\beta}$ and allow other aspects of data analysis to be done.

Example 7.2.3. *Suppose that we wish to take* $n = 30$ *observations overall. The approximate design in* (7.5) *may be converted to an exact* *design using Program_6 in the online repository. Running the program seven times produced three different allocations of observations to support points:* $(6, 9, 6, 9)$, $(7, 8, 6, 9)$ *and* $(7, 9, 5, 9)$. *I shall use the third of these, namely* $n_1 = 7$, $n_2 = 9$, $n_3 = 5$ *and* $n_4 = 9$. *Further suppose that* $y_1 = 2$, $y_2 = 2$, $y_3 = 4$ *and* $y_4 = 4$ *successes are observed at the four support points. The following program was run and the edited output appears below. [Recall that* **rep(a,n)** *means that a is repeated n times.]*

```
> x1 <- c(rep(-1,16),rep(0.9689,5),rep(1,9))
> x2 <- c(rep(-1,7),rep(1,14),rep(-1,9))
> y <- c(1,1,rep(0,5),1,1,rep(0,7),rep(1,4),0,rep(1,4),rep(0,5))
> out <- glm(y~x1+x2, family = binomial(link = "logit"))
> summary(out)

Call:
glm(formula = y ~ x1 + x2, family = binomial(link = "logit"))

Coefficients:
            Estimate Std. Error z value Pr(>|z|)
(Intercept)  -0.3846     0.3986  -0.965   0.3346
x1            0.7717     0.4217   1.830   0.0673 .
x2            0.3057     0.4192   0.729   0.4658
---
Signif. codes:  0 '***' 0.001 '**' 0.01 '*' 0.05 '.' 0.1 ' ' 1

(Dispersion parameter for binomial family taken to be 1)
```

To do further experimentation, one could refine the choice of values of $\boldsymbol{\beta}$ and their prior probabilities. Possible choices of β_i would be $\hat{\beta}_i \pm k_i \times \text{std.error}(\hat{\beta}_i)$, for some values of k_i. For example, use eight values of $\boldsymbol{\beta}$: $\left(\hat{\beta}_0 \pm k_0\,\text{std.error}(\hat{\beta}_0),\ \hat{\beta}_1 \pm k_1\,\text{std.error}(\hat{\beta}_1), \hat{\beta}_2 \pm k_2\,\text{std.error}(\hat{\beta}_2)\right)^\top$. A subject matter expert could allocate prior probabilities to the various values of $\boldsymbol{\beta}$.

7.2.3 Updating the probability distribution for $\boldsymbol{\beta}_1, \ldots, \boldsymbol{\beta}_h$

An alternative approach to that in Sub-section 7.2.2 is to retain the initial set of h parameter vectors $\boldsymbol{\beta}_1, \ldots, \boldsymbol{\beta}_h$ and to use the results of the experiment to update their associated prior probabilities. This would be done if one had considerable confidence in the proposed values of $\boldsymbol{\beta}_1, \ldots, \boldsymbol{\beta}_h$ and simply felt that the prior probabilities needed refining. Alternatively, you could use the estimate of $\boldsymbol{\beta}$ obtained from the experiment to guide the selection of a further design.

Updating the prior probabilities ψ_i for unchanged values of $\boldsymbol{\beta}_1, \ldots, \boldsymbol{\beta}_h$ is done using *Bayes' theorem,* also known as Bayes' Rule, which can be found in most introductory books on probability and statistics. It may be stated as follows:

Bayes' theorem:

Let A represent an event of interest, and let $B_1, \ldots, B_h$ represent a set of mutually exclusive and exhaustive events. (This means that exactly one of $B_1, \ldots, B_h$ must occur.) Then the conditional probability that event B_j occurs when we are told that A happens, denoted by $\Pr(B_j|A)$, *is given by*

$$\Pr(B_j|A) = \frac{\Pr(A|B_j)\Pr(B_j)}{\sum_{i=1}^{h}\Pr(A|B_i)\Pr(B_i)} \quad (j = 1, \ldots, h). \tag{7.6}$$

This statement of Bayes' theorem follows closely the notation used in Carlin & Louis (2009, p. 16).

In updating a Bayesian experimental design, let A represent the outcome from the experiment: that is, on taking n_i observations at the ith support point, $\boldsymbol{x}_i$, a response Y_i $(i = 1, \ldots, s)$ is obtained. The event B_j is the event that $\boldsymbol{\beta}_j$ is the true value of the parameter vector, and $\Pr(B_j)$ represents the prior probability that we have assigned to the event B_j; i.e., $\Pr(B_j) = \psi_j$. Thus (7.6) becomes

$$\Pr(B_j|A) = \frac{\Pr(A|B_j)\psi_j}{\sum_{i=1}^{h}\Pr(A|B_i)\psi_i} \quad (j = 1, \ldots, h).$$

Then $\Pr(B_j|A)$ is the posterior probability that $\boldsymbol{\beta}_j$ is the true value of $\boldsymbol{\beta}$ given that the outcome A was obtained in the experiment. This will be illustrated by an example.

Example 7.2.4. *Consider Example 7.2.3. For $m = 2$ and $h = 4$, B_j is the event that $\boldsymbol{\beta} = \boldsymbol{\beta}_j$, where $\boldsymbol{\beta}_1 = (-0.2, 0.8, 0.8)^\top$, $\boldsymbol{\beta}_2 = (-0.2, 1.2, 1.2)^\top$, $\boldsymbol{\beta}_3 = (0.2, 0.8, 1.2)^\top$ and $\boldsymbol{\beta}_4 = (0.2, 1.2, 0.8)^\top$, with prior probabilities $\psi_1 = 0.1$, $\psi_2 = 0.2$, $\psi_3 = 0.3$ and $\psi_4 = 0.4$, respectively. The numbers of observations at the four support points were $n_1 = 7$, $n_2 = 9$, $n_3 = 5$ and $n_4 = 9$, leading to $y_1 = 2$, $y_2 = 2$, $y_3 = 4$ and $y_4 = 4$ successes, respectively.*

Let p_{ij} represent the probability of a "success" on a single trial at the ith support point when the parameter vector $\boldsymbol{\beta}_j$ is used. Then

$$p_{ij} = 1/[1 + \exp(-\boldsymbol{f}^\top(\boldsymbol{x}_i)\boldsymbol{\beta}_j)], \quad i = 1, \ldots, s; \ j = 1, \ldots, h,$$

and the probability that y_i successes are observed in the n_i trials is

$$\Pr(Y_i = y_i|B_j) = \binom{n_i}{y_i} p_{ij}^{y_i}(1 - p_{ij})^{n_i - y_i}.$$

Recall that A is the event that y_i successes are observed from n_i trials at the ith support point for each $i = 1, \ldots, h$. As the results at different support points are assumed to be independent of one another, then

$$\begin{aligned}\Pr(A|B_j) &= \prod_{i=1}^{s} \Pr(Y_i = y_i|B_j)\\ &= \prod_{i=1}^{s} \binom{n_i}{y_i} p_{ij}^{y_i}(1-p_{ij})^{n_i-y_i} \quad (j = 1, \ldots, h).\end{aligned}$$

This result, together with the prior probabilities $\psi_j = \Pr(B_j)$, may be substituted into (7.6) and the posterior probabilities of $\boldsymbol{\beta}_1, \ldots, \boldsymbol{\beta}_4$ can be calculated.

Suppose that the experiment is conducted as described (i.e., $n_1 = 7$ observations at $\boldsymbol{x}_1 = (-1,-1)^\top$, ... , $n_4 = 9$ observations at $\boldsymbol{x}_4 = (1,-1)^\top$), and that $y_1 = 2$, $y_2 = 2$, $y_3 = 4$ and $y_4 = 4$ successes are observed. Then

$$\begin{aligned}p_{11} &= 1/[1+\exp(-\boldsymbol{f}^\top(\boldsymbol{x}_1)\boldsymbol{\beta}_1)]\\ &= 1/[1+\exp\{-(1,-1,-1)(-0.2,0.8,0.8)^\top\}]\\ &= 1/[1+\exp(1.8)]\\ &= 0.14185,\end{aligned}$$

$p_{21} = 1/[1+\exp\{-(1,-1,1)(-0.2,0.8,0.8)^\top\}] = 1/[1+\exp(0.2)] = 0.45017$, $p_{31} = 1/[1+\exp(-1.37512)] = 0.79821$ *and* $p_{41} = 1/[1+\exp(0.2)] = 0.45017$. *Note that $p_{11}, \ldots, p_{41}$ are the probabilities of a "success" at four different values of $\boldsymbol{x}$. There is no reason that the probabilities should sum to 1.*

So the event A is $\{Y_1 = 2, Y_2 = 2, Y_3 = 4, Y_4 = 4\}$ and

$$\begin{aligned}\Pr(A|B_1) &= \Pr(Y_1 = 2|B_1) \times \ldots \times P(Y_4 = 4|B_1)\\ &= \binom{7}{2} p_{11}^2(1-p_{11})^5 \times \binom{9}{2} p_{21}^2(1-p_{21})^7 \times \binom{5}{4} p_{31}^4(1-p_{31})^1\\ &\quad \times \binom{9}{4} p_{41}^4(1-p_{41})^5\\ &= 0.0.0023212629,\end{aligned}$$

and similarly $\Pr(A|B_2) = 0.0006746766$, $\Pr(A|B_3) = 0.0001785202$ *and* $\Pr(A|B_4) = 0.0008895905$. *Substitution of these values, and those of $P(B_i) = \psi_i$, into (7.6) gives the results $P(B_1|A) = 0.29895696$, $P(B_2|A) = 0.17378407$, $P(B_3|A) = 0.06897519$ and $P(B_4|A) = 0.45828377$.*

Calculation of the posterior probabilities can be done using Program_21 *in the online repository. Here* $\Pr(B_1|A), \ldots, \Pr(B_4|A)$ *are the conditional probabilities (given that the event* A *has occurred) that the four mutually exclusive and exhaustive events* $B_1, \ldots, B_4$ *occur, and the probabilities must sum to one.*

These posterior probabilities are quite different from the prior probabilities $\psi_1, \ldots, \psi_4$*. The value of each* ψ_i *may now be replaced by the new value* $P(B_i|A)$ *and the new values may be used to search for an updated Bayesian D-optimal design. Details of the calculations are not given as they follow those already displayed in Example 7.2.2. The Bayesian D-optimal design is found to be*

$$\xi^* = \left\{ \begin{array}{cccc} (-1,-1)^\top & (-1,1)^\top & (1,-1)^\top & (1,1)^\top \\ 0.210 & 0.294 & 0.293 & 0.203 \end{array} \right\}.$$

7.3 A continuous prior distribution for $\boldsymbol{\beta}$

7.3.1 Finding the D-optimal design

Sub-sections 7.2.1 to 7.2.3 considered the situation where the possible values postulated for each parameter formed a discrete set. For example, a researcher might choose the values -0.2 and 0.2 with equal probabilities of 0.5 as the prior distribution for β_0. However, an alternative approach is to postulate that the value of β_0 lies on an interval. For example, one might choose $\beta_0 \in \{\beta_0 : -0.2 < \beta_0 < 0.2\}$ as the domain for β_0. If there are $p = 3$ parameters, one might also choose $\beta_1 \in \{\beta_1 : 0 < \beta_1 < 1\}$ and $\beta_2 \in \{\beta_2 : 0 < \beta_2 < 2\}$.

Denote by $\mathcal{P}$ the *parameter space;* that is, the subset of $\mathbb{R}^p$ in which we believe that the parameter vector $\boldsymbol{\beta}$ lies. In the present example,

$$\mathcal{P} = \{(\beta_0, \beta_1, \beta_2)^\top : -0.2 < \beta_0 < 0.2,\ 0 < \beta_1 < 1,\ 0 < \beta_2 < 2\}. \quad (7.7)$$

Suppose that one has selected a continuous probability density function for the elements of $\boldsymbol{\beta}$ over $\mathcal{P}$. Denote this by $\psi(\boldsymbol{\beta})$. It is the continuous analogue of the discrete prior distribution that was given in Subsection 7.2. Instead of summing over the discrete values of $\boldsymbol{\beta}$ as was done in that sub-section, appropriate functions are *integrated* over the space $\mathcal{P}$. This gives the following multivariate integrals, analogous to (7.2) and (7.3), respectively, for the utility function and standardised variance:

$$\phi(\xi) = \int_{\mathcal{P}} \ln\left[\det\left(\boldsymbol{M}(\xi, \boldsymbol{\beta})\right)\right] \psi(\boldsymbol{\beta})\, d\boldsymbol{\beta} \quad (7.8)$$

and

$$d(\boldsymbol{x}, \xi^*) = \int_{\mathcal{P}} \omega(\boldsymbol{x}, \boldsymbol{\beta}) \boldsymbol{f}^\top(\boldsymbol{x}) \boldsymbol{M}^{-1}(\xi^*, \boldsymbol{\beta}) \boldsymbol{f}(\boldsymbol{x}) \psi(\boldsymbol{\beta})\, d\boldsymbol{\beta}. \quad (7.9)$$

Example: For $\mathcal{P}$ in (7.7), the utility function in (7.8) may also be written as

$$\phi(\xi) = \int_0^2 \int_0^1 \int_{-0.2}^{0.2} \ln\{\det[\boldsymbol{M}(\xi, \beta_0, \beta_1, \beta_2)]\}\, \psi(\beta_0, \beta_1, \beta_2)\, d\beta_0\, d\beta_1\, d\beta_2.$$

Except in trivial situations, it is not possible to find exact expressions for the values of the integrals in (7.8) and (7.9).

As the number of variables in a multiple integral increases, the number of evaluations of the function being integrated increases very considerably. In the integrals of (7.8) and (7.9), there are p variables (the p parameters in $\boldsymbol{\beta}$). The integral in (7.8) needs to be approximated for each candidate design considered in the process of maximising $\phi(\xi)$. The time required for the necessary computations using (say) Simpson's rule (see Section 2.5) quickly becomes prohibitive.

Researchers have sought alternative methods of numerical integration that require greatly reduced numbers of function evaluations. I will use results arising from Monahan & Genz (1997) and Gotwalt, Jones, & Steinberg (2009), who found methods to approximate the values of integrals of the form (7.8) when $\psi(\boldsymbol{\beta})$ is (i) a multivariate normal density function, or (ii) the probability density function for p independent variables β_i, the ith having a uniform distribution on the interval $[a_i, b_i]$. We will consider the second case.

The mathematics underlying the method is very much beyond the scope of this book, but a simple summary is that, for case (ii), we can write

$$\begin{aligned}\phi(\xi) &= \int_{\mathcal{P}} \ln[\det(\boldsymbol{M}(\xi, \boldsymbol{\beta}))]\, \psi(\boldsymbol{\beta})\, d\boldsymbol{\beta} \\ &\approx \sum_{i=1}^{N} w_i \ln[\det(\boldsymbol{M}(\xi, \boldsymbol{\beta_i}))] \qquad (7.10)\end{aligned}$$

and

$$d(\boldsymbol{x}, \xi^*) \approx \sum_{i=1}^{N} w_i\, \omega(\boldsymbol{x}, \boldsymbol{\beta}_i) \boldsymbol{f}^\top(\boldsymbol{x}) \boldsymbol{M}^{-1}(\xi^*, \boldsymbol{\beta}_i) \boldsymbol{f}(\boldsymbol{x}) \qquad (7.11)$$

for a comparatively small value of N using values of $w_1, \ldots, w_N$ and $\boldsymbol{\beta}_1, \ldots, \boldsymbol{\beta}_N$ that can be given to us by an R program. Equations (7.10) and (7.11) are very similar to (7.2) and (7.3), which were used for discrete prior distributions for the parameters. The problem for continuous prior distributions has been reduced to one for discrete distributions.

Program_22 in the online resources contains a function *RSquadrature.uniform* that will generate the weights w_i and abscissae $\boldsymbol{\beta}_i$ needed in

(7.10) and (7.11). The function calls on several other functions that are also given in Program_22. As well, the function *RSquadrature.uniform* needs a function, *halton*, that is in the package *randtoolbox* that is available on the CRAN website. On its first use, you will need to install *randtoolbox*. Subsequently, each time that you run R and need a function from this package, it will be necessary to issue the command

```
library(randtoolbox)
```

before running *RSquadrature.uniform*. The functions in Program_22 are incorporated in the R package *acebayes* (Overstall, Woods & Adamou, 2017), as described in Overstall & Woods (2017). I acknowledge with gratitude the generosity of Antony Overstall and David Woods in allowing them to be used.

The arguments for *RSquadrature.uniform* are the value of p, values of Nr and Nq (which alter the total number, N, of abscissae in the approximation), and a $p \times 2$ matrix *limits* that contains the limits for the uniform distributions for $\beta_1, \ldots, \beta_p$ (lower limits in column 1, upper limits in column 2). The output consists of an N-element vector of weights w_i, and an $N \times p$ matrix whose ith row contains the ith abscissa $\boldsymbol{\beta}_i$. These are used in (7.10) and (7.11).

Example 7.3.1. *Consider a logistic regression with* $\eta = \beta_0 + \beta_1 x_1 + \beta_2 x_2$ *(implying* $p = 3$ *and* $m = 2$*), and let the design space be* $\mathcal{X} = \{(x_1, x_2)\colon -1 \leq x_1 \leq 1, -1 \leq x_2 \leq 1\}$*. The prior distributions for the parameters are* $\beta_0 \sim U[-1,1]$*,* $\beta_1 \sim U[-3,3]$ *and* $\beta_2 \sim U[-3,3]$*. Program_23 below will generate the weights and abscissae for* (7.10) *and* (7.11).

```
p <- 3
Nr <- 3
Nq <- 4

#lower limits for parameters in column 1, upper limits in column 2
limits <- matrix(c(-1,-3,-3,1,3,3),p,2)

out <- RSquadrature.uniform(p, limits, Nr, Nq)
wts <- out$w
abscissae <- out$a
cbind(wts,abscissae)
  [1,] 1.523810e-01  0.0000000  0.00000000  0.00000000
  [2,] 1.266647e-02  0.0000000  1.46143658 -2.19215487
  [3,] 1.266647e-02 -0.8542417  0.62601014 -0.93901521
  [4,] 1.266647e-02 -0.6497583  1.57429131 -0.30493892
   :
   :
```

```
[241,] 3.602581e-05 -0.9999351  1.03599293 -1.55398939
```

There are 241 values of w_i *(in the first column of the output) and of* $\boldsymbol{\beta}_i$ *(in the rows spanning the second to fourth columns of the output). The values of* w_i *will be the same for two different runs of the program, but the values of* $\boldsymbol{\beta}_i$ *will vary from run to run. Consequently you will not necessarily get exactly the same answers from two evaluations of* (7.10) *or* (7.11).

This program used Nr = *3 and* Nq = *4, which are typical values to use. It is not advised that you decrease them, but you may increase them if you wish, although probably only by small amounts. (They must be integers.)*

Program_24 *directly follows* Program_23*, and uses the values of* p*,* Nr*,* Nq*,* wts *and* abscissae*. The program is very similar to other programs used earlier to search for optimal designs. The first thing necessary to decide is what value of* s *(the number of support points) to specify. The upper bound for non-Bayesian designs was* $p(p+1)/2$*, which is six here. However, it has already been mentioned (see page 69) that this upper bound does not apply for Bayesian experimental designs. I decided to try* $s = 10$. Program_24 *appears below:*

```
m <- 2
s <- 10
lim1 <- m*s
npts <- length(wts)

fx <- function(xvec)
{
 fvec <- c(1,xvec)
 fvec
}

detinfomatrix <- function(betavec,xvec,deswts)
{#calculates the determinant of the information matrix for
#a logistic regression in m variables.
#The input consists of the values of betavec, xvec and deswts
 xmat <- t(matrix(xvec,s,m))
 fxmat <- apply(xmat,2,fx)
 etavec <- as.vector(t(betavec)%*%fxmat)
 expetavec <- exp(etavec)
 modelwtvec <- expetavec/((1+expetavec)^2)
 infomat <- fxmat %*% diag(deswts*modelwtvec) %*% t(fxmat)
 det(infomat)
}

#calculate an approximation to the utility function
```

```
utility <- function(variables)
{
 xvec <- matrix(cos(pi*variables[1:lim1]),m,s,byrow=T)
 zvec <- variables[(lim1+1):((m+1)*s)]
 deswts <- zvec^2
 deswts <- deswts/sum(deswts)
 approx <- 0
 for (i in 1:npts)
 {
  temp <- detinfomatrix(abscissae[i,],xvec,deswts)
  approx <- approx + wts[i]*log(temp)
 }
 -approx
}

nsims <- 100
mindet <- 1000000
for (i in 1:nsims)
{
 initial <- c(runif(lim1),runif(s))
 out <- optim(initial,utility,NULL,method="Nelder-Mead")
 if(out$value < mindet) {mindet <- out$value
 bestdesign <- out$par
  }
 }
answer <- bestdesign
ansa <- matrix(cos(pi*answer[1:lim1]),m,s,byrow=T)
zvec <- answer[(lim1+1):((m+1)*s)]
deswts <- zvec^2
deswts <- deswts/sum(deswts)
solution <- rbind(ansa,deswts)
solution
mindet
```

This gave me the following results:

```
> solution
            [,1]        [,2]         [,3]        [,4]       [,5]
      -0.9987306  0.49410543  0.813505263 -0.20380411 -0.9999459
      -0.9956823 -0.96642698 -0.044382187  0.98666153  0.9694254
deswts  0.1547635  0.03001811  0.005799606  0.06232621  0.2171716
             [,6]          [,7]      [,8]       [,9]       [,10]
       0.12643142 -0.8839328280 0.9998190  0.9998968 -0.92863089
      -0.99802302 -0.5546368670 0.9968296 -0.8854396 -0.93142622
deswts  0.08822189  0.0003809722 0.2315761  0.1673043  0.04243773
> mindet
[1] 6.952569
```

I noted that support points 3 and 7 had very small design weights, so I decided to reduce the number of support points to $s = 8$ and ran the program again. The result was:

Design 7.1

```
> solution
             [,1]       [,2]        [,3]       [,4]        [,5]
       0.19219806  0.9588416 -0.96742452 -0.9979431  0.05633035
       0.99722036 -0.9934035  0.08369226 -0.9915185 -0.98810442
deswts 0.04173645  0.2071158  0.04759107  0.1915722  0.06426095
             [,6]       [,7]        [,8]
       0.99592141  0.9992664  -0.9883234
       0.22692883  0.9998860   0.9981643
deswts 0.08112878  0.1645251   0.2020696
> mindet
[1] 6.926874
```

I decided to base the starting values for the next search on the answers immediately above, and used the commands

```
oldx <- c(0.192,0.959,-0.967,-0.998,0.056,0.996,1,-0.988,0.997,
 -0.993,0.084,-0.992,-0.988,0.227,1,0.998)
olddel <- c(0.042,0.207,0.048,0.192,0.064,0.081,0.164,0.202)
startx <- acos(oldx)/pi
startdel <- sqrt(olddel/olddel[s])

mindet <- 1000000
for (i in 1:nsims)
{
 initial <- c(startx,startdel) + 0.1*(runif(lim1+s) - 0.5)
 out <- optim(initial,objective,NULL,method="Nelder-Mead")
 if(out$value < mindet) {mindet <- out$value
 bestdesign <- out$par
  }
 }
```

These commands back-transform the output values from the previous run to unconstrained values that can lie anywhere in $\mathbb{R}$. The first command inside the loop adds small positive or negative increments to each of these values, and then the first commands in the function utility transform the input to values that satisfy the required constraints.

After taking the output from the next lot of simulations, "tweaking" that output, and running the simulations again, the following output was obtained:

Design 7.2

```
> solution
              [,1]       [,2]        [,3]       [,4]        [,5]
       -0.04156896  1.0000000 -0.99996500 -0.9999940  0.05173776
```

```
        0.99999417 -0.9999997  0.07833031 -0.9999998 -0.99998262
deswts  0.07349709  0.1799858  0.07134747  0.1827745  0.07094532
             [,6]       [,7]       [,8]
         1.0000000  0.9999939  -0.9999913
        -0.0571310  0.9999954   0.9999878
deswts   0.0670904  0.1833768   0.1709826
> mindet
[1] 6.904864
```

Written another way, the suspected locally D-optimal design is

```
                    deltavec
[1,] -0.042  1.000     0.074
[2,]  1.000 -1.000     0.180
[3,] -1.000  0.078     0.071
[4,] -1.000 -1.000     0.183
[5,]  0.052 -1.000     0.071
[6,]  1.000 -0.057     0.067
[7,]  1.000  1.000     0.183
[8,] -1.000  1.000     0.171
```

Denote this design by ξ^*. The following commands come from Program_25 and calculate $M^{-1}(\xi^*, \beta_i)$ for each of the N (here $N = 241$) values of β generated by RSquadrature.uniform, and then store them in the array invinfomatrices.

```
infomat <- function(betavec,xvec,deswts)
{
 xmat <- t(matrix(xvec,s,m))
 fxmat <- apply(xmat,2,fx)
 etavec <- as.vector(t(betavec)%*%fxmat)
 expetavec <- exp(etavec)
 modelwtvec <- expetavec/((1+expetavec)^2)
 infomat <- fxmat %*% diag(deswts*modelwtvec) %*% t(fxmat)
 infomat
}

invinfomatrices <- rep(0,p*p*npts)
dim(invinfomatrices) <- c(p,p,npts)

optxvec <- c(-0.042,1,-1,-1,0.052,1,1,-1,1,-1,0.078,-1,-1,-0.057,1,1)
deltavec <- c(0.074,0.180,0.071,0.183,0.071,0.067,0.183,0.171)

for (h in 1:npts)
{
 betavec <- abscissae[h,]
 invinfomatrices[,,h] <- solve(infomat(betavec,optxvec,deltavec))
}
```

It is important to store these matrix inverses because this ensures that each inverse is calculated only once. The alternative is to calculate each matrix inverse every time that the standardised variance is calculated, which would be very wasteful from a computational perspective.

The remaining commands in Program_25 *calculate the standardised variance* $d(\boldsymbol{x}, \xi^*)$ *at each support point of Design 7.2, and then draw a contour plot of* $d(\boldsymbol{x}, \xi^*)$. *The standardised variance is equal to* $p = 3$ *at each support point, and* $d(\boldsymbol{x}, \xi^*) \leq 3$ *everywhere on the design space. (See Figure 7.3.) I am confident that Design 7.2 represents the Bayesian D-optimal design.*

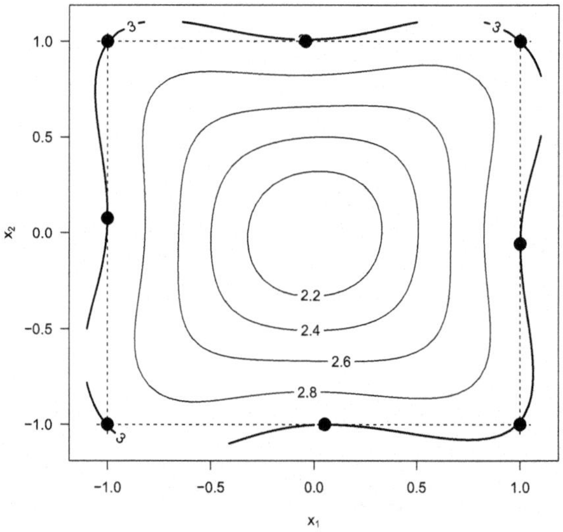

Figure 7.3 *Contour plot of the standardised variance for Design 7.2, the locally D-optimal design with* $s = 8$ *support points.*

The journey from the initial search for a design to the final conclusion that an optimal design has been obtained can be long and tortuous, as it depends very much on the initial starting point and the randomly generated alternatives to an intermediate answer. On another attempt to find the locally D-optimal design for the present situation, I quickly reached $s = 8$ *support points, and was then tempted by the low value of one design weight to try a design with* $s = 7$ *support points. This led to an apparent D-optimal design. However, a plot of the standardised variance, given in Figure 7.4, shows that its maximum value over the design space is greater than 3, so the design cannot be D-optimal. The plot strongly suggests that the support points should be approximately* $\boldsymbol{x} = (-1,-1)^\top$, $(-1,0)^\top$, $(-1,1)^\top$, $(0,-1)^\top$, $(0,1)^\top$, $(1,-1)^\top$, $(1,0)^\top$, $(1,1)^\top$*; i.e.,* $s = 8$ *points that are not too different from those of Design 7.2. This emphasises the importance of examining the standardised variance whenever searching for a D-optimal design.*

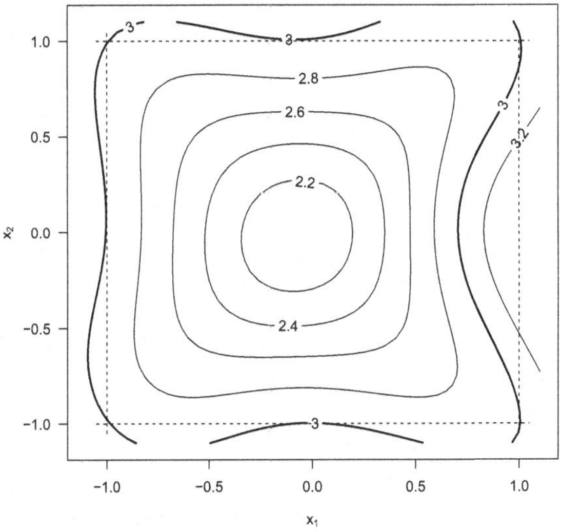

Figure 7.4 *Contour plot of the standardised variance for an apparent D-optimal design with* $s = 7$ *support points. Note that the value of the standardised variance exceeds* $p = 3$ *along most of the right-hand boundary.*

7.3.2 Using results from a previous experiment

Sub-section 7.2.2 described running an experiment and obtaining the results of a GLM analysis, including the parameter estimates and their estimated standard errors. In considering a discrete distribution for the values of the parameter vector $\boldsymbol{\beta}$ for use in finding a subsequent Bayesian D-optimal design, it was suggested that vectors of the form $\left(\hat{\beta}_0 \pm k_0\,\text{std. error}(\hat{\beta}_0), \ldots, \hat{\beta}_{p-1} \pm k_{p-1}\,\text{std. error}(\hat{\beta}_{p-1})\right)^\top$ be used, with appropriate prior probabilities assigned by a knowledgeable researcher. However, now a continuous distribution for $\boldsymbol{\beta}$ is being considered. In particular, it is assumed that the individual parameters are statistically independent of one another, and have uniform distributions.

A reasonable approach is to assume that the β_i are statistically independent, and that

$$\beta_i \sim U\left[\hat{\beta}_i - k_i \,\text{std. error}(\hat{\beta}_i),\ \hat{\beta}_i + k_i \,\text{std. error}(\hat{\beta}_i)\right] \quad (i = 0, 1, \ldots, p-1),$$

where the values of $k_0, \ldots, k_{p-1}$ are chosen by someone with knowledge of the experimental material. Having obtained a prior distribution for each parameter, one may then follow the example given in Subsection 7.3 that uses the results of the function *RSquadrature.uniform* to obtain a locally D-optimal Bayesian design.

The reader might prefer not to assume a "flat" (uniform) prior distribution for each parameter, but instead to select a distribution that has a peak at the ML estimate, and which tapers off on each side. normal distributions have such a shape, and the software is available if you wish to use a normal prior distribution for each parameter. However, if you would like another distribution (e.g., one with a triangular shape), I am not aware of any software that will allow you to do this and approximate the quantities in either (7.8) or (7.9).

The process to use normal prior distributions is fairly similar to that demonstrated in Example 7.3.1. The function that generates weights and abscissae to be used in numerical integration is called *RSquadrature.normal*, and is also made available by the kindness of Antony Overstall and David Woods. This program is stored as Program_26 in the Web site doeforglm.com. Like *RSquadrature.uniform*, three of the arguments for *RSquadrature.normal* are the values of p, Nr and Nq, which have the same meanings as they had in Example 7.3.1. The other two arguments for *RSquadrature.normal* are *mu* and *sigma*, respectively the mean vector and covariance matrix of the prior distribution of $\boldsymbol{\beta}$.

Example 7.3.2. *Let us pretend that the output in Sub-section 7.2.2 is what will be used to produce a prior distribution for* $\boldsymbol{\beta}$. *The relevant section of the output is*

```
Coefficients:
            Estimate Std. Error z value Pr(>|z|)
(Intercept)  -0.3846     0.3986  -0.965   0.3346
x1            0.7717     0.4217   1.830   0.0673 .
x2            0.3057     0.4192   0.729   0.4658
```

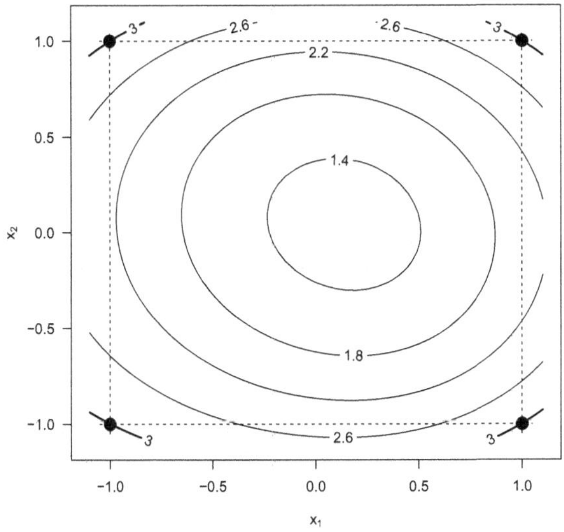

Figure 7.5 *A plot of the standardised variance $d(\boldsymbol{x}, \xi_6^*)$. Over the design space, $d(\boldsymbol{x}, \xi_6^*)$ achieves its maximum value of $p = 3$ at the four support points of ξ_6^*.*

If it is assumed that the parameters have independent distributions, then their covariances are zero, and it follows that

$$mu = \begin{bmatrix} -0.3846 \\ 0.7717 \\ 0.3057 \end{bmatrix} \quad and \quad sigma = \begin{bmatrix} 0.3986^2 & 0 & 0 \\ 0 & 0.4217^2 & 0 \\ 0 & 0 & 0.4192^2 \end{bmatrix}.$$

Provided that all necessary functions are present, a program that will search for a locally D-optimal design is very similar to the one on page 203, the only changes being the replacement of limits *by* mu *and* sigma *in the argument of the function, and using* RSquadrature.normal *instead of* RSquadrature.uniform. *Initially I used $s = 10$ support points, but was able to reduce it to $s = 4$. The locally D-optimal Bayesian design is*

$$\xi_6^* = \left\{ \begin{array}{cccc} (-1,-1)^\top & (-1,1)^\top & (1,-1)^\top & (1,1)^\top \\ 0.208 & 0.257 & 0.272 & 0.263 \end{array} \right\}.$$

The design's optimality was verified by a plot of the standardised variance, which appears in Figure 7.5.

Sub-section 7.2.3 used Bayes' theorem and the results of an experiment to obtain a posterior distribution for initially specified discrete values of $\boldsymbol{\beta}$. However, a warning was given that this should only be done if one had confidence in the initial values of $\boldsymbol{\beta}$ and wished to update the probabilities associated with those values. An equivalent procedure can be done when the prior distribution of $\boldsymbol{\beta}$ is continuous, but a similar warning is given here.

Let $\psi(\boldsymbol{\beta})$ and $f(\boldsymbol{y}|\boldsymbol{\beta})$ denote, respectively, the prior probability density function of $\boldsymbol{\beta}$ and the conditional distribution of the observations $\boldsymbol{y}$ for a given value of $\boldsymbol{\beta}$. The continuous analogue of (7.6) is

$$p(\boldsymbol{\beta}|\boldsymbol{y}) = \frac{f(\boldsymbol{y}|\boldsymbol{\beta})\psi(\boldsymbol{\beta})}{\int_{\mathcal{P}} f(\boldsymbol{y}|\boldsymbol{\beta})\psi(\boldsymbol{\beta})\, d\boldsymbol{\beta}}; \tag{7.12}$$

e.g., see Carlin & Louis (2009, p.15). The quantity $p(\boldsymbol{\beta}|\boldsymbol{y})$ is the posterior distribution of $\boldsymbol{\beta}$. An approximation of the denominator of (7.12) would again be calculated (for normal or uniform distributions) using the output of either *RSquadrature.normal* or *RSquadrature.uniform*.

If you do decide to use the result in (7.12), please consult a specialist in Bayesian procedures first. He or she may be able to suggest alternative computational procedures to help you, as there are several packages to help with calculations of this nature.

7.4 Exact Bayesian design

Recall that exact designs specify the support points of a design, and the design weight for each design is a multiple of $1/N$. A recent R package *acebayes* (Overstall, Woods & Adamou, 2017) will generate exact designs for designs of several variables and a large number of support points. The mathematics underlying the method is beyond the scope of this book, but the program is easy to apply. Install the package *acebayes* in the usual manner, and remember to issue the `library(acebayes)` command at the beginning of any R session that will use the program. The use of the package is illustrated by several examples:

Example 7.4.1. *Consider a logistic model in which the linear predictor* $\eta = \beta_0 + \beta_1 x_1 + \ldots + \beta_4 x_4$ *is a linear combination of* $m = 4$ *predictor variables, each assumed to lie on the interval* $[-1, 1]$*. The* $p = 5$ *parameters are assumed to have uniform prior distributions:* $\beta_0 \sim U[-3, 3]$, $\beta_1 \sim U[4, 10]$, $\beta_2 \sim U[5, 11]$, $\beta_3 \sim U[-6, 0]$ *and* $\beta_4 \sim U[-2.5, 3.5]$*. Suppose that we desire a Bayesian design of* $s = 6$ *support points with equal design weights* $\delta_1 = \ldots = \delta_6 = 1/6$*. The following program is based on an example given by Overstall & Woods (2017, Section 3.3). First, a seed is set so that you can duplicate this program. Then, the values of the parameters* s*,* p *and* m *are specified.*

```
> set.seed(1)
> n <- s <- 6
> p <- 5
> m <- 4
```

Create a starting matrix (a design) that has a row for each support point and a column for each predictor variable. Give a name to each variable. Each entry of the matrix has an entry that comes from the $U[-1, 1]$ *distribution. (The design represents a Latin hypercube sample — if you wish to know that!)*

```
> start.d1 <- matrix(2 * randomLHS(n=n,k=m) - 1,nrow=s,ncol=m,
 dimnames = list(as.character(1:n), c("x1", "x2", "x3", "x4")))
```

Vectors of the lower and upper bounds of the uniform prior distributions are specified. Then a function, prior1, is defined which will return a $B \times p$ *matrix, each of whose rows contains a sample of possible values of* $\beta_0, \ldots, \beta_{p-1}$.

```
> low <- c(-3, 4, 5, -6, -2.5)
> upp <- c(3, 10, 11, 0, 3.5)

> prior1 <- function(B){
 t(t(matrix(runif(n = p * B),ncol = p)%*%diag(upp-low,p)) + low)}
```

Use the function aceglm to generate the design. Arguments for the function include

- *the right-hand side of the formula specifying the linear model for* η,
- *the starting matrix,*
- *the distribution (the canonical link being assumed),*
- *specification of the name of the* $B \times p$ *matrix,*
- *the method to use in the calculations ("MC" for Monte Carlo, or the default of "quadrature"),*
- *the numbers of iterations in each of the first and second phases of the calculations (leave these as* $N1 = 1$ *and* $N2 = 0$ *unless you have read Overstall & Woods (2017) and know what you are doing),*
- *and the values of* B *for the two phases (leave these as c(1000, 1000) unless …).*

Typing `example1$phase2.d` *will give you the suggested design (s support points, each with design weight* $1/s$*).*

```
> example1 <- aceglm(formula=~x1+x2+x3+x4, start.d = start.d1,
 family = binomial, prior = prior1, method = "MC", N1 = 1,
 N2 = 0, B = c(1000, 1000))

> example1
```

```
Generalised Linear Model
Criterion = Bayesian D-optimality
Formula: ~x1 + x2 + x3 + x4
Family: binomial
Link function: logit

Method:  MC

B:  1000 1000

Number of runs = 6

Number of factors = 4

Number of Phase I iterations = 1

Number of Phase II iterations = 0

Computer time = 00:00:01

> example1$phase2.d
          x1          x2          x3         x4
1 -0.3571245  0.16069337 -0.61325375  0.9276443
2 -0.9167309  0.91411512  0.69842151  0.2605092
3 -0.8843699  0.42863930 -1.00000000 -0.9679402
4  0.3696224 -0.27126080  0.65284076  0.1850767
5  0.7172267 -0.34743402 -0.05968457 -0.6588896
6  0.7469636  0.05854029  1.00000000 -0.1742566
```

This second application of `aceglm` *defines the prior distribution through specification of the lower and upper limits of each uniform prior distribution, then uses the default method of "Quadrature" (as it is not specified) to find a design. The resulting design is printed out.*

```
> prior2 <- list(support = rbind(low, upp))

> example2 <- aceglm(formula = ~ x1 + x2 + x3 + x4,
  start.d = start.d1, family = binomial, prior = prior2,
  N1 = 1, N2 = 0)

> example2$phase2.d
          x1          x2         x3          x4
1 -0.3269814  0.08697755 -0.7583228  1.00000000
2 -0.8322237  0.86652194  0.5747066  0.51442169
3 -0.8987852  0.48881387 -0.8554894 -1.00000000
4  0.3441093 -0.29050147  0.4704248  0.07628932
5  0.8371670 -0.42361888  0.1429862 -0.95080251
6  0.6802119  0.10853163  1.0000000  0.75421678
```

The remaining two commands (below) perform 20,000 simulations of the weights and abscissae in order to use (7.10) *to estimate the utilities of the two designs that have been produced. Why do we do these simulations? It is because the values of* $\boldsymbol{\beta}$ *in* (7.10) *may be randomly generated (see the comment on page 204), resulting in the estimate of the utility function varying from simulation to simulation.*

Recall that we wish to find a design that maximises the utility. In this case, the design produced as example2 *would be selected, as its utility is greater.*

```
> mean(example1$utility(d = example1$phase2.d, B = 20000))
[1] -11.55139

mean(example2$utility(d = example2$phase2.d, B = 20000))
[1] -11.19838
```

Suppose that you wanted a design with $s = 8$ *equally weighted support points. Change the seed to 2 (if desired), replace* `s <- 6` *by* `s <- 8`*, and run the whole program again. In a matter of a second or so, you obtain the following output.*

```
> example1$phase2.d
           x1          x2         x3         x4
1 -0.57935559  0.48889067 -0.1252828  1.0000000
2  0.36068797 -0.17574732  1.0000000 -0.5815204
3 -0.87202905  0.47172578 -0.8397567 -1.0000000
4  1.00000000 -1.00000000  0.2536735  0.9275464
5  0.48711174 -0.51398955 -1.0000000 -1.0000000
6 -0.02739129 -0.30128404 -0.3181002  1.0000000
7  0.29561906  0.24367197  0.9547421 -0.3048681
8 -0.82325960  0.01640478 -1.0000000 -1.0000000

> prior2 <- list(support = rbind(low, upp))
>
> example2 <- aceglm(formula = ~ x1 + x2 + x3 + x4,
  + start.d = start.d, family = binomial, prior = prior2,
  + N1 = 1, N2 = 0)


>
> example2$phase2.d
           x1          x2           x3         x4
1 -0.59098376  0.49543197 -0.063874594  1.0000000
2  0.36815759 -0.12925704  1.000000000 -0.5050266
3 -0.88789143  0.50487774 -0.964392629 -1.0000000
4  0.97268747 -0.95919978 -0.008130879  0.2815559
5  0.16463809 -0.51398955 -1.000000000 -1.0000000
6  0.01527268 -0.27162117  0.169227869  1.0000000
7  0.33383668  0.23257864  0.706116963 -0.3855793
```

```
8 -0.81859188  0.02673269 -0.737791431 -1.0000000

> mean(example1$utility(d = example1$phase2.d, B = 20000))
[1] -8.629215

> mean(example2$utility(d = example2$phase2.d, B = 20000))
[1] -8.769524
>
```

A comparison of the two average utilities suggests that the first of the two designs is to be preferred.

This description of the use of *aceglm* presents it like a "black box." This is because the underlying theory is beyond the scope of this book. However, I trust that the example makes it clear how to use the function. You need only specify the values of s, p, m, *low* and *upp*, the names of the m variables, and the family (e.g., binomial), and you can run the program exactly as it is. If you have the theoretical background to understand Overstall & Woods (2017) and the references mentioned therein, you should certainly use the program. If not, then be aware that the method provides an extremely fast way to generate a Bayesian design by two methods when each parameter has a uniform prior distribution and the distributions of the parameters are statistically independent. It does *not* give an approximate design that can be shown to be Bayesian D-optimal, but often it will give a design that is clearly good, as the following example shows.

Example 7.4.2. *Recall Example 5.3.5, where a locally D-optimal design was found for a Poisson regression with* $m = 4$, $p = 5$, $s = 5$ *and* $\boldsymbol{\beta} = (1, 2, 1, -1, -2)^\top$. *The support points have equal design weights. If instead we consider a Bayesian design, and choose prior distributions* $\beta_0 \sim U[0.8, 1.2]$, $\beta_1 \sim U[1.6, 2.4]$, $\beta_2 \sim U[0.8, 1.2]$, $\beta_3 \sim U[-1.3, -0.7]$ *and* $\beta_4 \sim U[-2.2, -1.8]$, *which restrict the values of the parameters to be close to those in* $\boldsymbol{\beta} = (1, 2, 1, -1, -2)^\top$, *what design does* aceglm *yield?*

```
> example1$phase2.d
          x1         x2         x3          x4
1  1.00000000  0.8999860 -1.0000000 -1.00000000
2  1.00000000 -0.9114759 -1.0000000 -1.00000000
3  0.97299720  1.0000000 -1.0000000 -0.02606028
4  1.00000000  1.0000000  0.8572299 -1.00000000
5 -0.05462328  1.0000000 -1.0000000 -1.00000000

> example2$phase2.d
          x1         x2         x3          x4
1 -0.3258347  0.8401790  0.2923183  0.73158736
2 -0.1692818 -0.8846324 -0.9234802 -0.03759992
```

```
3 -0.8442510 -0.8376870  0.6645709 -1.00000000
4 -0.1520863 -0.9490834 -0.5390256  0.89899266
5  0.8892828 -1.0000000  1.0000000  0.28930294

> mean(example1$utility(d = example1$phase2.d, B = 20000))
[1] 29.48662

> mean(example2$utility(d = example2$phase2.d, B = 20000))
[1] -4.370936
```

The design from example1 *would be preferred over that from* example2 *because of a greater mean utility function. The selected example agrees very well with the locally D-optimal design from Example 5.3.5, suggesting that — in a comparable situation —* acebayes *will give an acceptable design.*

7.5 Final comments

Bayesian experimental design removes from researchers the burden of having to design an experiment for a GLM with the design being crucially dependent on what may be a very unreliable guess of the value of the parameter vector, $\boldsymbol{\beta}$. Being able to provide an interval of possible values, and an associated probability distribution, for each parameter eases that dependence.

Theoretical and computational developments in Bayesian experimental design are progressing rapidly. It seems very likely that Bayesian experimental design will continue to become more common, and convenient. You are encouraged to monitor the literature for new developments.

Bibliography

ALBERT, A. & ATKINSON, J.A. (1984). On the existence of maximum likelihood estimates in logistic regression models. *Biometrika* **71**, 1–10.

ATKINSON, A.C. & DONEV, A.N. (1992). *Optimum Experimental Designs.* Oxford University Press.

ATKINSON, A.C., DONEV, A.N. & TOBIAS, R.D. (2007). *Optimum Experimental Designs, with SAS.* Oxford University Press.

BOX, G.E.P., HUNTER, J.S. & HUNTER, W.G. (2005). *Statistics for Experimenters: Design, Innovation and Discovery.* Hoboken, NJ: John Wiley & Sons, 2nd edn.

CARLIN, B.P. & LOUIS, T.A. (2009). *Bayesian Methods for Data Analysis.* Boca Raton, FL: Chapman & Hall, 3rd edn.

CHALONER, K. & LARNTZ, K. (1989). Optimal Bayesian designs applied to logistic regression experiments. *Journal of Statistical Planning and Inference* **21**, 191–208.

CHALONER, K. & VERDINELLI, I. (1995). Bayesian experimental design: a review. *Statistical Science* **10**, 273–304.

CRAWLEY, M.J. (2013). *The R Book.* Chichester, UK: John Wiley & Sons, 2nd edn.

DE MICHEAUX, P.L., DROUILHET, R. & LIQUET, B. (2013). *The R Software: Fundamentals of Programming and Statistical Analysis.* New York: Springer.

DOBSON, A.J. & BARNETT, A.G. (2008). *An Introduction to Generalized Linear Models.* Boca Raton, FL: Chapman & Hall/CRC Press, 3rd edn.

DROR, H.A. & STEINBERG, D.M. (2006). Robust experimental design for multivariate generalized linear models. *Technometrics* **48**, 520–529.

FARAWAY, J.J. (2006). *Extending the Linear Model with R: Generalized Linear, Mixed Effects and Nonparametric Regression Models.* Boca Raton, FL: Chapman & Hall/CRC Press.

FIRTH, D. (1993). Bias reduction of maximum likelihood estimates. *Biometrika* **80**, 27–38.

FORD, I., TORSNEY, B. & WU, C.F.J. (1992). The use of a canonical form in the construction of locally optimal designs for non-linear problems. *Journal of the Royal Statistical Society B* **54**, 569–583.

GOTWALT, C.M., JONES, B.A. & STEINBERG, D.A. (2009). Fast computation of designs robust to parameter uncertainty for nonlinear settings. *Technometrics* **51**, 88–95.

GUTTMAN, I. (1982). *Linear Models: An Introduction.* New York: Wiley.

HARVILLE, D.A. (1997). *Matrix Algebra From a Statistician's Perspective.* New York: Springer.

KIEFER, J. & WOLFOWITZ, J. (1960). The equivalence of two extremum problems. *Canadian Journal of Mathematics* **12**, 363–366.

KOSMIDIS, I. & FIRTH, D. (2009). Bias reduction in exponential family nonlinear models. *Biometrika* **96**, 793–804.

KUEHL, R.O. (2000). *Design of Experiments: Statistical Principles of Research Design and Analysis.* Pacific Grove, CA: Brooks/Cole, 2nd edn.

MCCULLAGH, P. & NELDER, J.A. (1989). *Generalized Linear Models.* London, UK: Chapman & Hall, 2nd edn.

MONAHAN, J. & GENZ, A. (1997). Spherical-radial integration rules for Bayesian computation. *Journal of the American Statistical Association* **92**, 664–674.

NELDER, J.A. & MEAD, R. (1965). A simplex method for function minimization. *The Computer Journal* **7**, 308–313.

NELDER, J.A. & WEDDERBURN, R.W.M. (1972). Generalized linear models. *Journal of the Royal Statistical Society, Series A* **135**, 370–384.

OVERSTALL, A.M. & WOODS, D.C. (2017). Bayesian design of experiments using Approximate Coordinate Exchange. *Technometrics* **59**, 458–470.

OVERSTALL, A.M., WOODS, D.C. & ADAMOU, M. (2017). acebayes: an R package for Bayesian optimal design of experiments via Approximate Coordinate Exchange. https://arxiv.org/abs/1705.08096. (accessed 12th August, 2018).

PUKELSHEIM, F. (1993). *Optimal Design of Experiments.* New York: John Wiley & Sons.

R CORE TEAM (2018). *R: A Language and Environment for Statistical Computing.* R Foundation for Statistical Computing, Vienna, Austria. URL `https://www.R-project.org/`.

ROCKAFELLAR, R.T. (1970). *Convex Analysis.* Princeton, NJ: Princeton University Press.

Russell, K.G., Eccleston, J.A., Lewis, S.M. & Woods, D.C. (2009a). Design considerations for small experiments and simple logistic regression. *Journal of Statistical Computation and Simulation* **79**, 81–91.

Russell, K.G., Woods, D.C., Lewis, S.M. & Eccleston, J.A. (2009b). D-optimal designs for Poisson regression models. *Statistica Sinica* **19**, 721–730.

Searle, S.R. (1971). *Linear Models.* New York: Wiley.

Searle, S.R. (1982). *Matrix Algebra Useful for Statistics.* New York: Wiley.

Silvey, S.D. (1980). *Optimal Design: An Introduction to the Theory for Parameter Estimation.* London: Chapman & Hall.

Thompson, G.P. (2010). Optimal design for generalized linear models with a multinomial response. Ph.D. thesis, School of Mathematics and Applied Statistics – Faculty of Informatics, University of Wollongong.

UCLA: Statistical Consulting Group (2015). What is complete or quasi-complete separation in logistic/probit regression and how do we deal with them? `http://www.ats.ucla.edu/stat/mult_pkg/faq/general/complete_separation_logit_models.htm`. (accessed 5th August, 2016).

Wikimedia (2018). List of statistical packages. `https://en.wikipedia.org/wiki/List_of_statistical_packages`. (accessed 8th February, 2018).

Woods, D.C., Lewis, S.M., Eccleston, J.A. & Russell, K.G. (2006). Designs for Generalized Linear Models with several variables and model uncertainty. *Technometrics* **48**, 284–292.

Zhang, Y. (2006). Bayesian D-optimal design for generalized linear models. Ph.D. thesis, Virginia Polytechnic Institute and State University.

Zocchi, S.S. & Atkinson, A.C. (1999). Optimum experimental designs for multinomial logistic models. *Biometrics* **55**, 437–444.

Index

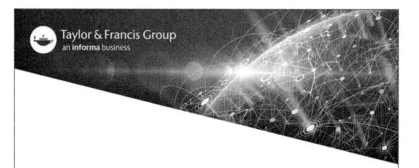

Printed in the United States
by Baker & Taylor Publisher Services